Fundamentals of Test Measurement Instrumentation

Fundamentals of Test Measurement Instrumentation

by Keith R. Cheatle

Copyright © 2006 by   ISA – Instrumentation, Systems, and Automation Society
                             67 Alexander Drive
                             P.O. Box 12277
                             Research Triangle Park, NC 27709

All rights reserved.

Printed in the United States of America.
10 9 8 7 6 5 4 3 2

ISBN-13:   978-1-55617-914-3
ISBN-10:   1-55617-914-6

No part of this work may be reproduced, stored in a retrieval system, or transmitted in any form or by any means, electronic, mechanical, photocopying, recording or otherwise, without the prior written permission of the publisher.

Notice

The information presented in this publication is for the general education of the reader. Because neither the author nor the publisher has any control over the use of the information by the reader, both the author and the publisher disclaim any and all liability of any kind arising out of such use. The reader is expected to exercise sound professional judgment in using any of the information presented in a particular application.

Additionally, neither the author nor the publisher have investigated or considered the effect of any patents on the ability of the reader to use any of the information in a particular application. The reader is responsible for reviewing any possible patents that may affect any particular use of the information presented.

Any references to commercial products in the work are cited as examples only. Neither the author nor the publisher endorses any referenced commercial product. Any trademarks or tradenames referenced belong to the respective owner of the mark or name. Neither the author nor the publisher makes any representation regarding the availability of any referenced commercial product at any time. The manufacturer's instructions on use of any commercial product must be followed at all times, even if in conflict with the information in this publication.

Library of Congress Cataloging-in-Publication Data

Cheatle, Keith.
  Fundamentals of test measurement instrumentation / by Keith Cheatle.
    p. cm.
  Includes index.
  ISBN 1-55617-914-6 (pbk.)
 1. Electronic instruments. 2. Testing--Equipment and supplies. 3. Physical measurements. I. Title.
  TK7878.4.C43 2006
  620'.0044--dc22
                              2006000524

Acknowledgments

The creation of this book has been a rewarding experience that was made easier through the efforts of a number of others.

I would like to acknowledge the encouragement and support of my wife, Barb, who has been steadfastly behind the project from the start. Her efforts to share knowledge of her own chosen profession were an inspiration to me.

Our two Shi Tzus, Cricket and Misty, provided many hours of company sleeping below the computer desk while I hit the keyboard above.

I would also like to acknowledge the helpful comments of the reviewer that certainly improved the content of the book.

Finally I would like to acknowledge the work of the ISA editorial staff whose expert advice made the book much easier to read.

About the Author

Keith R. Cheatle is a Professional Engineer registered in the Province of Ontario, Canada. He holds a Bachelor's degree from Ryerson Polytechnical Institute (now Ryerson University) in Toronto. He joined ISA in 1965 as a student member, became a member in 1966, a senior member in 1982, and a life member in 1998.

Mr. Cheatle worked for over 37 years in the Instrumentation Section of Atomic Energy of Canada Limited and was Section Head for 25 years. This section provided test instrumentation systems for use in a development laboratory environment. It tested components and systems that were to be used in nuclear power stations. The section was also responsible for the process instrumentation used on the high pressure, high temperature circulating water test loops used to simulate nuclear reactor operating conditions.

In addition, the Instrumentation Section operated a state-of-the-art calibration laboratory for in-house calibration of instrumentation systems under an ISO 9001 quality assurance program. Mr. Cheatle was responsible for the design and operation of this calibration facility.

As the Section Head, Mr. Cheatle trained technicians and junior engineers in the application of test instrumentation. This training consisted of both lectures and on-the-job instruction.

Table of Contents

About the Author vii

Foreword xv

1 Introduction to Test Measurement Instrumentation 1

    1.1   Introduction 1
    1.2   Why and Where Test Measurement Instrumentation Is Used 2
    1.3   Components of a Test Measurement System 3
    1.4   Characteristics of an Ideal Test Measurement System 10
    1.5   Differences between Test Measurement Instrumentation and Process Instrumentation 14
    1.6   Smart Transducers 16

2 Static Transducer Characteristics 19

    2.1   Hysteresis 26
    2.2   Deadband 29
    2.3   Linearity 32
    2.4   Conformity 39
    2.5   Repeatability 39
    2.6   Accuracy 41
    2.7   Resolution 42
    2.8   Environmental Effects on Transducer Characteristics 43

3 Dynamic Transducer Characteristics 45

    3.1   Frequency Response 47
    3.2   Response Time 49
    3.3   Damping 50

4 Transducer Types 53

    4.1   Temperature Transducers 53
    4.2   Pressure Transducers 67
    4.3   Flow Transducers 77
    4.4   Displacement Transducers 89
    4.5   Velocity Transducers 99
    4.6   Acceleration Transducers 100
    4.7   Force Transducers 105
    4.8   Strain Transducers 108

5 Signal Conditioning 117

    5.1   Voltage Amplifiers 119
    5.2   Linearizing Amplifiers 122
    5.3   Strain Gauge Amplifiers 123
    5.4   Charge Amplifiers 126
    5.5   Filters 128
    5.6   Signal Isolators 129
    5.7   Analog-to-Digital Converters 130
    5.8   Differential versus Single-Ended Inputs 131

6 Transducer Installation 135

    6.1   Wiring and Cabling 136
    6.2   Grounding 138
    6.3   Connecting to the Test Rig 138

7 Data Acquisition 143

    7.1   Digital Indicators 143
    7.2   Recorders 146
    7.3   Data Loggers 146
    7.4   Data Acquisition Systems 147
    7.5   Filters 151

8 Data Reduction and Analysis 153

    8.1   Graphing 154
    8.2   Sorting 155
    8.3   Signal Analyzers 156

9 Equipment Calibration. 159

    9.1   Component Calibration 165
    9.2   System Calibration 166
    9.3   In Situ Calibration 166
    9.4   Calibration Standards 167
    9.5   Multifunction Calibrators 168
    9.6   Calibration Records 169
    9.7   Calibration Laboratory Requirements 171
    9.8   Calibration System Software 172
    9.9   Software Validation 174
    9.10   Quality Assurance System Requirements 175
    9.11   International, National, and Other Standards 176
    9.12   Calibration Equipment Specifications 178

10 Pressure Calibration 183

    10.1   Dead-Weight Testers 183
    10.2   Water Dead-Weight Testers 189
    10.3   Oil Dead-Weight Testers 190
    10.4   Pneumatic Dead-Weight Testers 190
    10.5   Digital Calibrators 191
    10.6   Manometers 193

11 Temperature Calibration 197

    11.1   Temperature Calibration Baths 198
    11.2   Dry Block Calibrators 203
    11.3   Temperature Calibration Furnaces 204
    11.4   Infrared Calibrators 205
    11.5   Thermocouple and RTD Simulators 206
    11.6   Standard Platinum Resistance Thermometers 208
    11.7   Triple Point of Water Cells 210
    11.8   Metal Fixed Point Cells 212
    11.9   Ice Baths 214
    11.10 Boiling Point of Water 215

12 Electrical Calibration 217

    12.1   Voltage 217
    12.2   Current 219
    12.3   Resistance 220
    12.4   Thermocouples 222
    12.5   Multifunction Calibrators 223

13 Force Calibration 227

    13.1   Dead-Weight Force Calibration 227
    13.2   Proving Rings 233
    13.3   Hydraulic Presses 234
    13.4   Shunt Calibration 235

14 Flow Calibration 239

    14.1   Provers 241
    14.2   Weigh Tank 243
    14.3   PVTt Calibration Facility 245
    14.4   Reference Flow Sensor 246

15 Displacement Calibration 249

    15.1   Gauge Blocks 250
    15.2   Dial Indicators 251

16 Vibration Calibration 253

    16.1   Shaker Table 260
    16.2   Reference Accelerometer 262

17 Strain Gauge Calibration 263

    17.1   Bridge Shunt Calibration 264

18 System Accuracy 269

    18.1   System Accuracy Statistics 270
    18.2   Root Sum Square Versus Sum of Errors 275
    18.3   Standards Accuracy 277
    18.4   Test Uncertainty Ratio 278
    18.5   Uncertainty Versus Accuracy 279

19 Transducer Specifications 285

    19.1   Interpreting Manufacturer's Specification Sheets 286
    19.2   Transducer Type 297
    19.3   Operating Principle 298
    19.4   Accuracy (% of full scale vs. % of reading) 298
    19.5   Linearity 299
    19.6   Hysteresis 300
    19.7   Environmental Operating Limits 300

20 Test Procedure Design 303

    20.1   Test Objective 304
    20.2   Required Measurements 304
    20.3   Accuracy Requirements 304
    20.4   Frequency Response Requirements 305
    20.5   Transducer Test Environment Limits 306
    20.6   Transducer Connections and Space Limitations 306
    20.7   Test Data Format 307
    20.8   Test Data Analysis 308
    20.9   Transducer Selection 309
    20.10 Calibration Requirements 309

Appendix 1
    Internet Links to Measurement Instrumentation Information.....311

Appendix 2
    Internet Links to General Engineering Information.............323

Index 325

Foreword

This book introduces you to the exciting world of test measurement instrumentation. Extensive coverage is provided about the transducers, signal conditioning, data acquisition, and data analysis equipment used to make typical measurements. These measurements are routinely made in research and development laboratories, product development laboratories, and manufacturing facilities. The automotive industry and the consumer electronics industry are including more and more measurement systems in order to improve their products.

You are informed about how the separate components mentioned above are combined to form a measurement system and about the advantages and disadvantages of the various transducer types. The description of static and dynamic characteristics of transducers leads you into an understanding of the factors that affect the accuracy of the system and into the concept of measurement uncertainty. Detailed information about the calibration of measurement systems and the equipment used to carry out calibrations is a major part of the book. A comprehensive listing of Internet links to further test measurement information and links to related general engineering information is provided in appendices.

With a few exceptions, the subject of measurement instrumentation is not taught as part of technician and engineering curriculums in colleges and universities. Knowledge of the subject is typically gained through years of on-the-job experience and attendance at courses designed to gain knowledge of specific equipment or individual parts of the wide-ranging subject. The book strives to provide the basic information required to start making meaningful measurements and hopefully minimize the mistakes that can be made by improper application of the equipment.

Working in the field of test measurement instrumentation is both challenging and rewarding. One is able to work with leading-edge technology and able to apply knowledge of many engineering disciplines. The work is constantly changing and constantly requiring the practitioner to learn new theoretical and practical information. I hope that the contents of this book will inspire you to join the ranks of the measurement instrumentation professionals.

Keith R. Cheatle, P.Eng.

1

Introduction to Test Measurement Instrumentation

"I often say that when you can measure what you are speaking about, and express it in numbers, you know something about it, but when you cannot express it in numbers, your knowledge is of a meagre and unsatisfactory kind; it may be the beginning of knowledge, but you have scarcely, in your thoughts, advanced to the stage of Science, whatever the matter may be."
— Sir William Thompson, Lord Kelvin (1827–1907)

1.1 Introduction

The opening quotation, by Lord Kelvin, shows the importance of being able to measure parameters and record or display the resulting measurements in numbers. Measurements allow us to make specific conclusions about the physical conditions of a test object and about how the object responds to changes in its environment. These measurements allow us to verify that the object will perform correctly under the specified conditions and verify that future changes to the object do not affect its performance. The quotation also underscores the important role that measurements play in advancing scientific knowledge.

This book introduces students, technicians, technologists, and engineers to the specialized test instrumentation that is used in R&D laboratories, in testing organizations, and in industrial maintenance departments. The book will focus on the practical application of test instrumentation and emphasize the importance of creating a "measurement system" that involves components, installation, wiring, and calibration. The design, application, and calibration of systems for measuring pressure, temperature, flow, force, displacement, and vibration will be discussed. A significant portion of the book will be devoted to the calibration of test instrumentation, including detailed information about calibration equipment, calibration methods, and calibration records.

You will learn to:

- design test measurement systems
- select appropriate equipment
- understand the characteristics of system components
- understand system and component calibration
- understand the operating principles of transducers
- determine overall accuracy of a system
- formulate basic test procedure designs

1.2 Why and Where Test Measurement Instrumentation Is Used

Test measurement instrumentation is used to obtain the values of the physical parameters that exist in an environment, process, or product. Examples of these physical parameters include temperature, pressure, flow, motion, force, and strain. We need to make these measurements in order to test theories and equipment, to verify that a product's specifications are met, to monitor the operation of equipment, or simply to tell us how hot it is outside. Test measurement instrumentation is used in research and development laboratories, in product development centers, in manufacturing plants, in maintenance facilities, and in commercial and household products.

Research and development (R&D) laboratories are prime users of measurement instrumentation. Pure research that expands knowledge and verifies the theoretical advances in science is carried out by government labs, university labs, and the labs of large companies. The development laboratories apply the theoretical knowledge and scientific advances to improve the equipment, operation, and facilities used by society. Product development centers then design, test, and improve new consumer and industrial products based on the work of the R&D laboratories. Manufacturing plants produce the final products and use measurement instrumentation to verify that their products meet specifications. This measurement activity takes place during the production and maintenance phases of the plants' work. Measurement instrumentation is increasingly being built into commercial and household products. Automobiles are now full of sensors and displays, for example, and appliances have ever more sophisticated measurements and controls.

1.3 Components of a Test Measurement System

A test measurement system usually consists of several separate components for measuring some physical parameter that are joined by some means of transferring the measurement information about the parameter between the components until the information is finally presented in a form that the system operator can use. The components that are found in a typical system include the sensor, the signal conditioner, and the display. Most electrical and electronic-based systems use copper wiring as a means of transferring measurement information while other systems can use pneumatic, hydraulic, optical, or wireless transmission.

In its simplest form, such as a pressure gauge, the components of a test measurement system are all contained within the same device. The sensing element (bourdon tube, bellows, diaphragm, etc.) moves in a manner that is proportional to the applied pressure. This movement is transferred to the gauge pointer through a series of gears, levers, and cams, which translate the element motion into a proportional rotational movement that drives the gauge pointer. The pointer moves to a position that is equivalent to the applied pressure, and the resulting pressure is read by the user on a circular scale located under the pointer and attached to the face of the gauge. In the case of the pressure gauge, the bourdon tube, bellows, or diaphragm is the transducer or sensor. The gears, levers, and cams form the transmission system; the pointer and dial form the display or readout.

An example of a slightly more complex test measurement system is a digital pressure gauge. In this case, the sensing element, such as a dia-

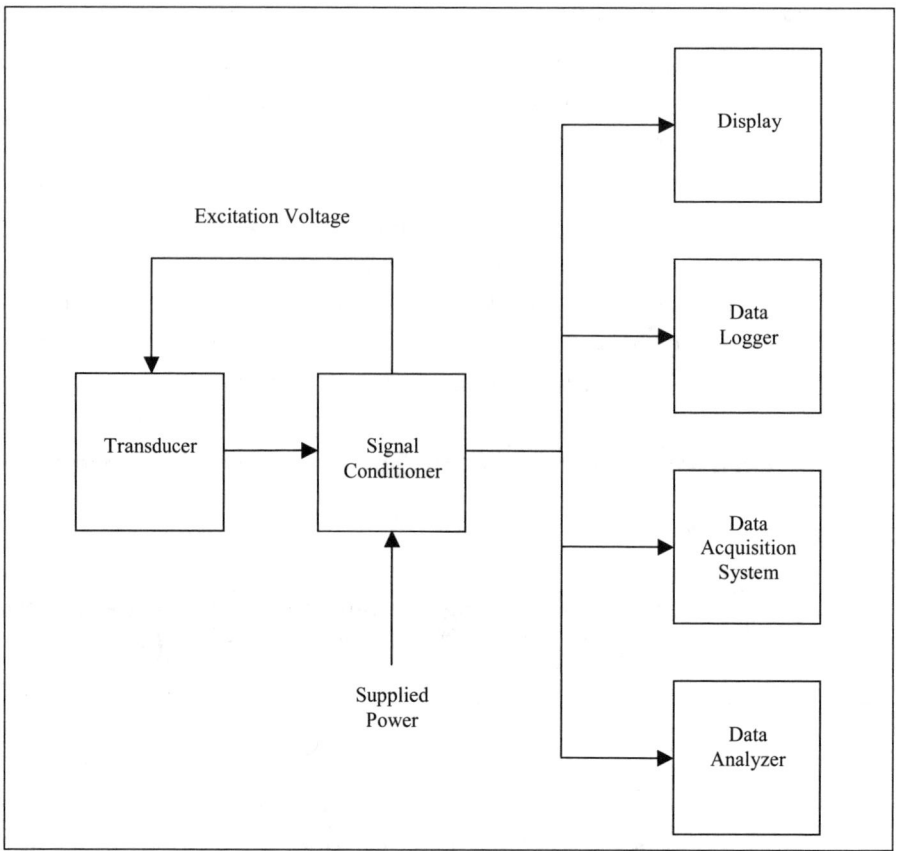

Figure 1-1. Typical Test Measurement System

phragm, has strain gauges attached that respond to the strain in the diaphragm material that occurs as a result of the applied pressure. Wiring connects the strain gauges to a power source and to an electronic circuit that converts the resistance change of the strain gauges into an analog voltage. Other electronics are used to linearize the voltage and provide adjustments that scale the voltage so it is proportional to the applied pressure. The analog voltage is then converted into an equivalent digital voltage, and this voltage is used to drive a digital display, which indicates the pressure in the desired units. The digital voltage can also be used to drive a liquid crystal display that mimics the traditional pressure gauge pointer and scale.

Additional electronics can provide a voltage or current output that can connect the gauge with wires to other equipment so the pressure reading can be used for remote displays, data recording, or the control of other equipment. A power source such as a battery or DC power supply must be included in the system in order to provide power for the electronics and displays. This electronics can include a microprocessor, which is used to accept the raw electronic signal and perform the data manipulation needed to provide display and output signals. In the case of a digital pressure gauge, the transducer or sensor is the diaphragm with its attached strain gauges. The sensor output and sensor power are transmitted through wires to the electronic signal conditioning circuits, and the user reads the pressure on the liquid crystal display on the gauge.

An example of a complex test measurement system is the measurement of the exterior surface temperature of a space vehicle. The temperature sensor must be mounted in such a way that it is protected during the launch, space, and re-entry environments (potentially subjecting it to vibration, vacuum, radiation, space debris) while ensuring that it is still close enough to the exterior surface to make an accurate temperature measurement. The sensor's wiring is led through the space vehicle's wall in such a way that the pressure containment between the vehicle's interior and the space environment is preserved. The wiring is connected to a signal conditioning unit that provides the excitation power for the sensor and electronic circuits so the sensor's signal from the sensor can be processed into a form of signal that the measurement system can use. This signal could be an analog voltage, a digital voltage, a current, or a digital signal to be connected to a computer bus.

The temperature signal could be routed to displays for the crew, to computers for data storage, and to on-board alarm systems to warn crew of abnormal conditions. The signal could also be sent by radio transmission to satellites that would then retransmit the temperature data to one or more ground stations. This would enable ground personnel to monitor the vehicle's exterior temperature.

1.3.1 Transducers

Transducers or sensors are the front end of the measurement system. They are the devices that interface directly with the equipment or environment about which we want information. They transform the physical parameter of interest (pressure, temperature, vibration, etc.) into a form that can be further used and processed by other devices until the user eventually has the information he or she desires.

Some transducers, such as thermocouples or piezoelectric accelerometers, have a self-generating output. In this case, the sensor element provides an electrical output through the action of the physical parameter on the sensor element. The thermocouple provides a small voltage output whenever the temperature at the junction of the two joined metal wires is different from that at the ends of the wires where the voltage is read with a voltmeter. The piezoelectric sensor element provides an electrical charge output whenever a change of force is made to the crystal sensor element.

Other transducers such as resistance temperature detectors (RTDs) and strain-gauge-based transducers need a power supply in order to produce an electrical signal. In the case of the RTD, a small electrical current is passed through the resistive element so the signal- conditioning circuit can determine the resistance of the RTD and thus the temperature associated with that resistance. In the strain gauge transducer, one or more resistive strain gauges are connected in a Wheatstone bridge circuit. This Wheatstone bridge circuit operates by the application of an excitation voltage to it, which enables the resistance change of the strain gauges to be translated into an equivalent voltage change.

There are thousands of transducers available commercially today, and hundreds more are introduced every year. Each transducer has a specific design that makes it more or less suitable for a given application. It is the job of the test measurement professional to select the most appropriate sensor for the application at hand. This means he or she must understand the basic operating principle of the transducer, the complete specifications of the transducer, the test environment and range of test conditions, the measurement requirements, the effect of external influences on the transducer, and the transducer's installation and calibration requirements. It is the goal of this book to give you information about a wide range of sensors so this selection process will be easier. With experience and knowledge, you will learn that certain transducer types are more suitable to certain measurement situations, simplifying the selection process.

1.3.2 Signal Conditioners

Signal conditioners lie at the center of the measurement system. They serve to receive the output from the transducer, carry out some basic processing of the signal, and pass the signal on to the display, storage, and analysis components. Signal conditioners may also supply an excitation voltage to power the transducer. Some transducers are self-contained units that have the signal conditioning circuit built into the body of the transducer. In strain gauge pressure transducers, the Wheatstone bridge completion resistors, temperature compensating circuits, regulated bridge power supply, and amplifier can all be contained in the transducer. The only external connections are an unregulated power supply and an output display. The advantages of this type of transducer include less electrical noise pickup on the output signal, less expensive cabling, and ease of use. The disadvantages include limitations on the environment in which the transducer can be used (temperature, pressure, radiation, etc.) and nonrepairable electronics, which means plants may have to discard this more expensive type of transducer when it is damaged.

The number and variety of signal conditioners is smaller than that of transducers. Certain types of signal conditioners can be used with a number of transducers that share a common operating principle. For example, a strain gauge signal conditioner can be used with a strain gauge pressure transducer, a strain gauge load cell, or with strain gauges used to determine stresses in materials. A charge amplifier can be used with piezoelectric accelerometers or piezoelectric pressure transducers. Each type of signal conditioner, however, is still designed to perform to certain specifications and to have certain features. The test measurement professional must still ensure that the signal conditioner is compatible with the selected transducer and that the signal conditioner's specifications meet the test measurement requirements.

1.3.3 Displays

Displays are the simplest form of system output device used to measure the output from the signal conditioning circuit. They indicate to the user the value of the parameter being applied to the transducer. Displays can be analog or digital meters that indicate the raw voltage from the signal conditioner, or they can be meters that scale the raw voltage so as to provide an output in units of measurement (°C or psig). Other displays include bar graphs and alarm indicators or a combination of both. These displays can be panel mounted or bench mounted, or they can be portable hand-held units. Most of these displays rely on the user to read and note the value being measured, although some displays have some limited ability to store and retrieve data. Most displays are designed to measure static or very slowly changing data since these are the only types of signals that users can process visually. An oscilloscope, on the other hand, is an example of a display that is designed for time varying or dynamic data.

1.3.4 Data Loggers

Data loggers are standalone display systems that are used to store data from a large number of sensors. Data loggers are usually used to record data that is changing fairly slowly. A snapshot of all the data channels is taken at approximately the same time, and the data is stored along with the time at which it was taken. The scan is repeated periodically, from a few times a second to a few times an hour. The data is later retrieved from the logger's memory device and presented to the user in tabular or graphic form. Tabular outputs present the data in a table, in which the data from all or selected channels is included along with its time of acquisition. Graphical displays provide the measurement information in the form of an X-Y graph, where the x-axis shows the acquisition time and the y-axis shows the multiple data points from one or more channels. The distinction between data loggers and computer-based data acquisition systems is somewhat blurry, and today the latter have largely replaced the standalone data logger.

1.3.5 Data Acquisition Systems

A data acquisition system is a refinement of the data logger but with the additional flexibility that a computer-based system with software provides. The software lets the user customize the way the system acquires, stores, and displays the test data. Multiple sensors are connected to the input of the data acquisition system. The outputs of the sensors are sequentially scanned by a multiplexer, which switches each signal to a common processing circuit. This common circuit can consist of filters, analog-to-digital converters, or even a common signal conditioning circuit. Once the data is in a digital format, it can be stored, manipulated, and displayed in any number of ways. Data acquisition systems are capable of accessing hundreds of signals at speeds of over a million samples per second. They are ideally suited to measuring parameters that are changing very quickly.

Data acquisition systems range in size from portable, hand-held devices that access only a few points to personal computer–based systems and even to systems controlled by mainframe computers.

1.3.6 Data Analyzers

Data analyzers are a specialized version of data acquisition systems that have additional software that processes and analyzes the sensor data. Analyzers can be designed for a specific type of measurement such as vibration analysis, or they can be general-purpose analyzers capable of manipulating data in a variety of formats. Analyzers that use the Fourier analysis technique to deal with data in the frequency and time domains are an example of the general-purpose analyzer. These analyzers can be portable and capable of accessing only a few channels, or they can be computer-based systems that access a large number of channels. The portable systems as shown in Figure 1-2 have an oscilloscope-type display for real-time display of the raw and analyzed data. Computer-based systems have monitors for the data display. A typical computer-based system is shown in Figure 1-3.

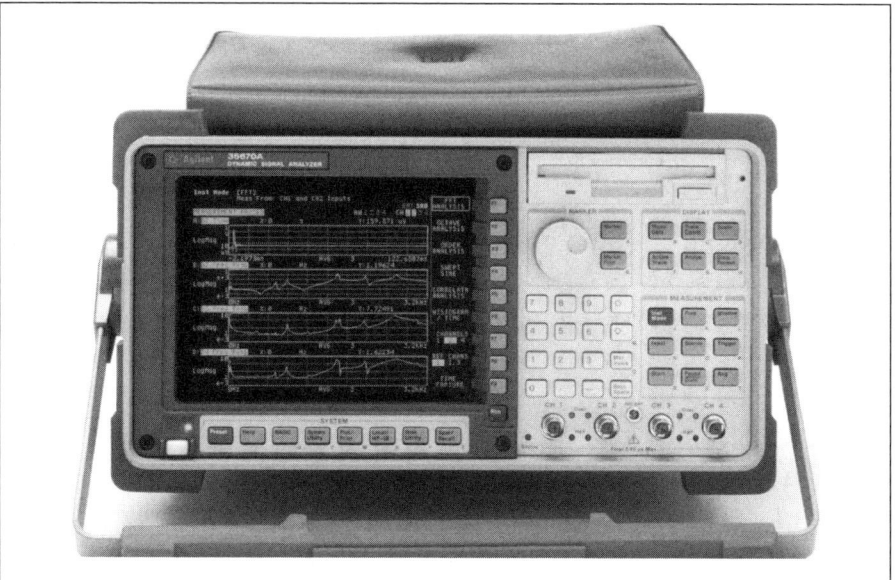

Figure 1-2. Portable Data Analyzer
(© Agilent Technologies, Inc. 2000
Reproduced with Permission, Courtesy of Agilent Technologies, Inc.)

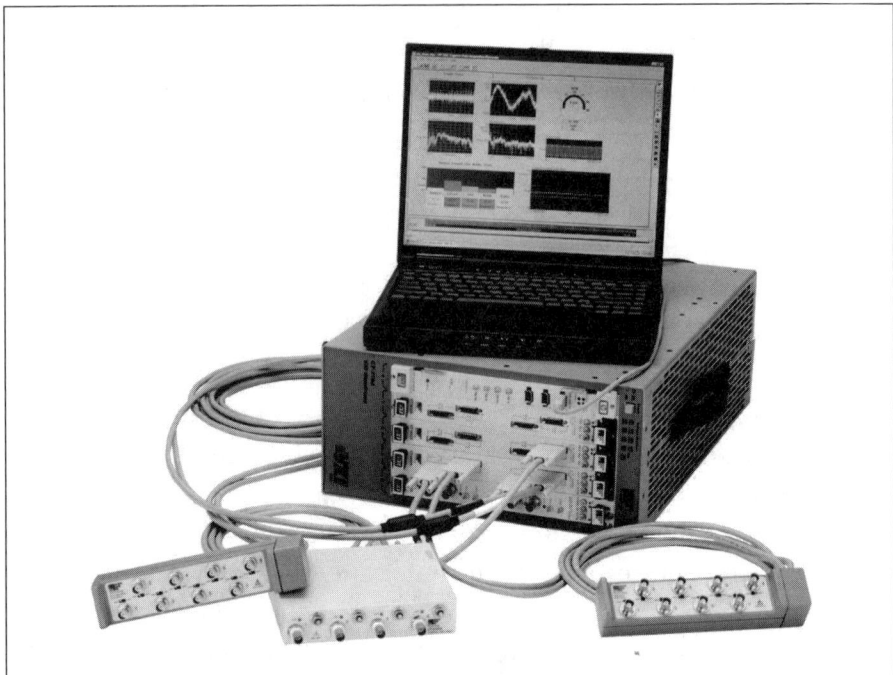

Figure 1-3. Computer-based Multi-channel Data Analyzer
(Courtesy, VXI Technology)

1.4 Characteristics of an Ideal Test Measurement System

Although the characteristics of an ideal test measurement system can depend on the total system, including sensor, cabling, signal conditioning, and displays, most of the characteristics of an ideal test measurement system involve the sensor or transducer. This is because it is the sensor that is located in the test environment or mounted to the test specimen. The rest of the system can be located remotely from the environment and specimen in a more benign area. This remoteness also ensures that the equipment does not have any effect on the test environment or on the test specimen. Such effects could mask or alter the behavior of the specimen in a way that impacts the test results to the point that erroneous conclusions are made.

The ideal sensor should have the following characteristics:

- *Be physically small and have a low mass.* Small size is important since the sensor often has to be attached to an existing piece of equipment which poses its space and accessibility limitations.

Tests are usually carried out on real or prototype equipment that was not designed to have sensors attached once the tests are completed. Small size is also important in ensuring that the sensor does not significantly change the test environment. The sensor's size can affect the flow distribution of the fluid surrounding the test specimen, for example. This changing flow pattern could influence the vibration behavior of the test specimen or affect the heat transfer between specimen and fluid.

Low mass is essential to ensure that the mass of the specimen with the sensor attached is almost the same as the mass without the sensor. The added mass will affect the vibration characteristics of the specimen and could also affect the heat transfer between specimen and surrounding atmosphere. If we wanted to know the acceleration of a specimen when exposed to a specific force exerted by the system that is connected to the specimen, we would mount an accelerometer on the specimen and measure the acceleration. The acceleration is equal to the force divided by the mass of the specimen plus the attached accelerometer. If the accelerometer mass is 1% of the mass of the specimen then the measured acceleration will be 1% lower than it would have been if the accelerometer was not present.

- *Not be influenced by extremes in the environmental conditions, such as temperature, pressure, radiation, and water or chemicals, to which it is exposed during the test.* The sensor should be immune to changes in these conditions as well as to the absolute level of these conditions. In reality, every sensor has environmental limitations, and the test measurement professional must select a sensor that can withstand the environmental test conditions. He or she must also understand and evaluate the impact that the magnitude and change in these conditions can have on the measurement system's output.

- *Be capable of measuring a wide range of variation in the value of the measured parameter, be highly accurate, and be highly linear.* Wide rangeability enables one sensor to cover the entire range of test conditions imposed on the specimen. This reduces the cost of having to install multiple sensors, reduces the space required by multiple sensors, and reduces the number of tests that have to be run when sensors that fail under test conditions have to be replaced. High sensor accuracy reduces the measurement uncertainty and increases the confidence that the effect of the measured parameter on the specimen can be determined and reported. High sensor linearity improves both rangeability and accuracy. It also simplifies data analysis since a simple linear equation can be used in the data reduction process.

- *Be capable of a wide frequency response.* The sensor should be able to respond equally well to static or quickly changing conditions in the measurement parameter. The transducer's installation method, the wiring, and the frequency response capabilities of the rest of the measurement system also contribute to the system's overall frequency response. The operating principle of some types of sensors prevents them from being used under static or very slowly changing measurement parameter conditions. Piezoelectric sensors, for example, only provide an output when the measurement parameter is changing faster than a given threshold level. At the other extreme, all sensors have some limit to the upper frequency they can respond to. This is determined by factors such as sensor mass, physical construction, material limits, electronics, resonant frequencies, and so on.

Other characteristics of the ideal sensor that need no further explanation are low cost, high output signal, ruggedness, ease of operation, and ease of attachment.

Since the ideal sensor does not exist, some compromise must be made when selecting which sensor to use in a given situation. The important thing for the test measurement professional is to identify and understand the limitations of the selected sensor. He or she should also account for the effects of these limitations on the uncertainty of the measurement. The magnitude of each of the limitations is provided by reputable sensor manufacturers and is spelled out in the sensor specification sheet. The user must determine the effects of any conditions not covered by the specification sheet. This may require the user to carry out tests on the sensor before using it in the planned test.

1.4.1 Process Influence on Transducer

The transducer's operating characteristics will be affected by the process to which it is attached and by the environment surrounding it. Conditions such as ambient temperature, pressure, and radiation will affect the transducer's specifications. The environmental operation limits to which the transducer has been designed will be documented in the specification sheet. The magnitude of the effect that each environmental condition has on certain specifications within the environmental limits may also be documented in the specifications. For example, the upper temperature limit for a pressure transducer may be specified as 250°F. The manufacturer has designed this transducer to operate up to this temperature. Use above this limit could damage the transducer to the point that it would no longer operate. The effect on the transducer's accuracy caused by using it at a temperature above room temperature may be ±0.01% per °F. At a room temperature of 70°F, the transducer's accuracy may be ±1%. If the transducer is operated at a temperature of 170°F, the accuracy would be changed by (170 − 70) × 0.01% = 1%, for a total accuracy of 1% + 1% = 2%.

The process to which the transducer is attached can also affect the transducer and its specifications. Unless a transducer is designed for underwater use, it should not be submerged, for example. If the transducer is to be connected to a fluid that is corrosive then the materials of construction must be compatible with the fluid.

1.4.2 Transducer Influence on Process

As described in section 1.4, sensor or transducer characteristics such as size and mass can affect the environment we are trying to measure or affect the performance of the equipment we are testing. One example of this influence includes using long steel-sheathed thermocouples that protrude from the test environment and conduct heat from the test fluid into the surrounding atmosphere (this is called stem conduction). Another is using resistance temperature detectors in a poorly conductive fluid or gas. The heat generated by the current passing through the resistance element of the RTD (this is called self-heating) is added to the heat produced by the fluid or gas. This can cause the temperature measurement to be slightly higher than if the self-heating was removed by a highly conductive fluid or gas. Selecting a transducer that minimizes the influence on the process is part of the job of the test measurement professional. He or she must also understand and determine the degree to which measurements are affected by using a certain transducer.

1.4.3 Invasive versus Noninvasive Systems

A measurement system that has a large impact on the process or equipment being tested is described as an invasive system. Using a linear variable differential transformer (LVDT) to measure the dynamic displacement of a piece of equipment could be invasive, for example. The fact that the LVDT core must contact or be connected to the equipment changes the vibration characteristics of the equipment. The physical size of the core could change the flow patterns of the fluid around the equipment.

A noninvasive measurement system is one that has a very small or nonexistent effect on the equipment being measured. Using a laser displacement transducer would eliminate the problems just mentioned since there is no physical contact between the equipment and the transducer, and the laser beam would not affect the flow distribution.

Some compromise must be made when deciding which transducer to use since the cost of a laser displacement system may be much greater than that of an LVDT-based system. The deciding factor is whether the uncertainty in the validity of the measurement caused by using an invasive system is too large. The degree of uncertainty is something that must be decided before selecting the transducer system. The test measurement professional must determine how much uncertainty can be tolerated while still meeting the requirements of the test.

1.5 Differences between Test Measurement Instrumentation and Process Instrumentation

Test measurement instrumentation includes the transducers, signal conditioners, and displays used in laboratory, manufacturing, and maintenance organizations to measure physical parameters. These components are required during research and development testing, product development, production evaluation, and maintenance and repair operations. Figure 1-4 illustrates a typical test measurement pressure transducer. This pressure transducer is approximately three eighths of an inch long and weighs 0.2 grams.

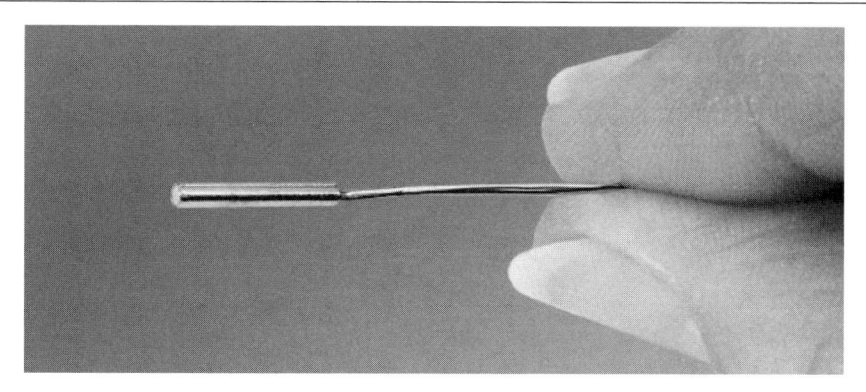

Figure 1-4. Test Measurement Pressure Transducer
(Courtesy, Kulite Semiconductor Products, Inc.)

Process instrumentation is the class of measurement equipment used in industrial process industries such as refineries, power plants, and chemical producers. It is used to measure, control, and provide alarm and shutdown signals. There is some overlap in these descriptions since process industries may also use test instrumentation, and test organizations can use process instrumentation if it suits the application. Table 1-1 provides the major differences between test measurement instrumentation and process instrumentation.

Table 1-1. Differences Between Test and Process Instrumentation

Test Measurement Instrumentation	Process Instrumentation
Used in laboratories, product development	Used in process industries
Designed for a specific application	Designed for a wide range of applications
Small and lightweight	Large and heavy
High frequency response	Static to low frequency response
Less stable	More stable
Less rugged	More rugged
Usually more accurate	Usually less accurate
Expensive due to smaller quantities used	Less expensive
Usually has a voltage output	4–20 mA DC output
Wider environmental range	More limited in environmental range
Typically uses separate signal conditioning	Self-contained signal conditioning
Sensors referred to as transducers	Sensors referred to as transmitters

The process instrument transmitter shown in Figure 1-5 is approximately nine inches high and weighs approximately twelve pounds.

Figure 1-5. Process Instrumentation Pressure Transmitter
(Courtesy, Emerson Process Management, Rosemount, Inc.)

1.6 Smart Transducers

An Institute of Electrical and Electronics Engineers (IEEE) standard has been developed so analog output sensors can be easily interfaced with compatible signal conditioning and data acquisition hardware. IEEE 1451.4, Mixed Mode Interface for Smart Transducers, defines a method for providing a digital serial interface that downloads a digital file known as a Transducer Electronic Data Sheet (TEDS). This file contains information about the transducer that is read by the interfacing hardware, which then configures itself to the correct setup. This setup was previously set manually through hardware or software to provide the proper configuration. This effectively produces a plug-and-play type of system that relieves the user of much of the work associated with the system setup.

The TEDS file contains information about the manufacturer, model number, serial number, measurement range, sensitivity, and calibration data. This file can reside in an Electrically Erasable Programmable Read Only Memory (EEPROM) chip within the sensor body or in a file within the data acquisition hardware, the data acquisition computer, or even in a web-accessible file. This Virtual TEDS allows the TEDS concept to be applied to older sensors that do not have a built-in EEPROM or to sensors that require remote storage because of the transducer's extreme operating environment. The sensor's analog output is maintained as the means for providing the output signal from the transducer. This combination of analog and digital signals gives rise to the description "mixed mode."

The TEDS file follows a standard format that can be used by the compatible interfacing hardware. The file is divided into three sections as follows:

1. The first section of the TEDS file contains the basic identifying information for the sensor such as manufacturer, model number, and serial number.

2. The second section of the TEDS file provides the sensor's specific data sheet information. This information allows the system to be configured and to convert the measurement signal into engineering units. This data can include measurement range, electrical output range, sensitivity, power requirements, and calibration data. The IEEE 1451.4 standard specifies a number of "Standard TEDS," known as templates that have been created to define most of current types of sensors in use today. These include templates for accelerometers, microphones, pressure transducers, Wheatstone bridge sensors, strain gauges, force sensors, thermocouples, RTDs, thermistors, LVDTs, resistive sensors, and sensors with current or voltage outputs. The standard also allows custom templates to be created to cover special parameters and requirements.

3. The third section of the TEDS file is for use by the operator for any custom information associated with a particular sensor. This information can include such things as sensor location and calibration due date.

The IEEE 1451.4 standard also defines two types of digital interfaces that are used to transmit the TEDS data. The Class I interface is used by transducers that have a constant current powered piezoelectric sensing element (IEPE transducers), such as accelerometers and microphones. The TEDS data is transmitted while the current is switched in one direction, and the measurement data is transmitted over the same pair of wires while the current is switched in the other direction. The Class II interface is used for most types of transducers and requires a separate pair of wires in order to transmit the digital TEDS data.

2
Static Transducer Characteristics

Static transducer characteristics are determined by applying a series of non–time varying signals to the input of the transducer and noting the corresponding outputs. Each transducer is designed by its manufacturer to make measurements over a certain range of input values. The manufacturer will state the upper and lower limits of this range in the specification sheet.

A pressure transducer, for example, may be designed to measure from 0 psig to 100 psig. In this case, the range is 0 to 100 psig. The lower range limit is 0 psig, while the upper range limit is 100 psig. Operating the transducer beyond these range limits may damage the transducer or permanently change one or more of the specifications. The manufacturer may build in mechanical stops that limit the amount of overrange, thus preventing permanent damage to the transducer. Other manufacturers will state the amount of overrange that can be tolerated before permanent damage occurs. The transducer will have a stated output signal range that is proportional to the input range. The output range for the pressure transducer that has a range of 0 to 100 psig may be 0 to 100 millivolts DC. In this case, the output from the transducer will be 0 millivolts when the input is 0 psig and 100 millivolts when the input is 100 psig. If this was an ideal, linear transducer, then the output values for intermediate input values would lie on a straight line between 0 millivolts and 100 millivolts. Thus, an input of 20 psig would produce an output of 20 millivolts, and an input of 80 psig would produce an output of 80 millivolts. The span of a transducer is defined as the algebraic difference between the upper range limit and the lower range limit. In this case, the span is (100 psig – 0 psig) = 100 psig.

The transducer or the associated signal conditioning will be able to adjust the output signal so it is exactly equal to or as close as possible to the ideal 0 millivolts when the input is 0 psig and to 100 millivolts when the input is 100 psig. These adjustments are named the "zero adjustment" and the "span adjustment," respectively. The zero adjustment sets the output signal to 0 millivolts when the input is 0 psig, while the span adjustment sets the output signal to 100 millivolts when the input is 100 psig. The span adjustment changes the slope of the output-versus-input line or the gain. In most cases, the adjustment is made by means of a potentiometer (variable resistor). In microprocessor-based transducers, the adjustment may be made by zero and span pushbuttons that are activated at the zero and full-scale input settings.

In many transducers the zero and span adjustments are interactive. Adjusting the span potentiometer will usually affect the zero output; however, adjusting the zero potentiometer will usually not affect the span. This makes the adjustment process iterative, or back and forth. The zero is first set, followed by the span. Adjusting the span affects the zero, which then must be reset. The span is then rechecked and adjusted as needed. This procedure is repeated as many times as necessary until the output signals are exactly as required.

Some transducers also have an adjustment that will minimize the nonlinearity of the transducer. In other cases, the manufacturer optimizes the transducer for minimal nonlinearity, and no adjustment is provided.

Adjusting the transducer output by applying known inputs is called "transducer calibration." The purpose of such calibration is to ensure that the transducer is as close as possible to the ideal transducer.

The data shown in Table 2-1 contains the results of a calibration of a pressure transducer that has an input range of 0 to 100 psig and a corresponding output range of 0 to 100 millivolts. The transducer was able to be adjusted so that the real output was exactly the same as the ideal output. The error column in Table 2-1 shows the difference between the real output and the ideal output (error = real output − ideal output). In this case, there is no error in the output of the transducer. The two final columns in the table show two other ways that the transducer error can be expressed. The first is to express the error as a percentage of the ideal full-scale output (% FS). In this case, the ideal full-scale output is 100 millivolts, and the error as a percentage of full scale is calculated as (Error/100) × 100. The second method is to express the error as a percentage of the ideal reading (% Reading). When the reading is 10 millivolts, the error is calculated as (Error/10) × 100. When the reading is 20 millivolts, the error is calculated as (Error/20) × 100. This is continued in the same manner for each reading. Note that an error at a reading of 0 millivolts is not calculated since the formula would contain a division by zero, which would result in an undefined number. An error at a reading very close to 0 millivolts (0.01 millivolts) could be calculated and used in place of the 0 millivolts reading.

Table 2-1. Calibration Results for an Ideal Transducer

Input psig	Ideal Output millivolts	Real Output millivolts	Error millivolts	Error % FS %	Error % Reading %
0	0	0	0	0	0
10	10	10	0	0	0
20	20	20	0	0	0
30	30	30	0	0	0
40	40	40	0	0	0
50	50	50	0	0	0
60	60	60	0	0	0
70	70	70	0	0	0
80	80	80	0	0	0
90	90	90	0	0	0
100	100	100	0	0	0

Transducer manufacturers prefer to state the error as a percentage of full scale (% FS) since this results in a smaller-magnitude number. The transducer user is probably more interested in the error as a percentage of reading since this determines the accuracy of the measurement at any part of the range. A transducer with a range of 0 to 100 psig and an error specification of ±1% FS can have this error occur at any point in the range. If the error occurs at 100 psig, then the error in the reading would be ±1%. If the error occurs at 50 psig, then the error in the reading would be ±2%. If the error occurs at 20 psig then the error in the reading would be ±5%. This shows how important it is to select a transducer whose upper-range limit is close to the value of the measurement. It also shows how important it is to calibrate a transducer so we know where the maximum errors occur in the range.

Figure 2-1 shows the calibration data from Table 2-1 in a graphical format. The input range of 0 to 100 psig is graphed on the *x*-axis while the output range of 0 to 100 millivolts is graphed on the *y*-axis. In this case, the transducer's real output is exactly equal to the ideal output so the line through the real output data points lie directly on top of the ideal output data points.

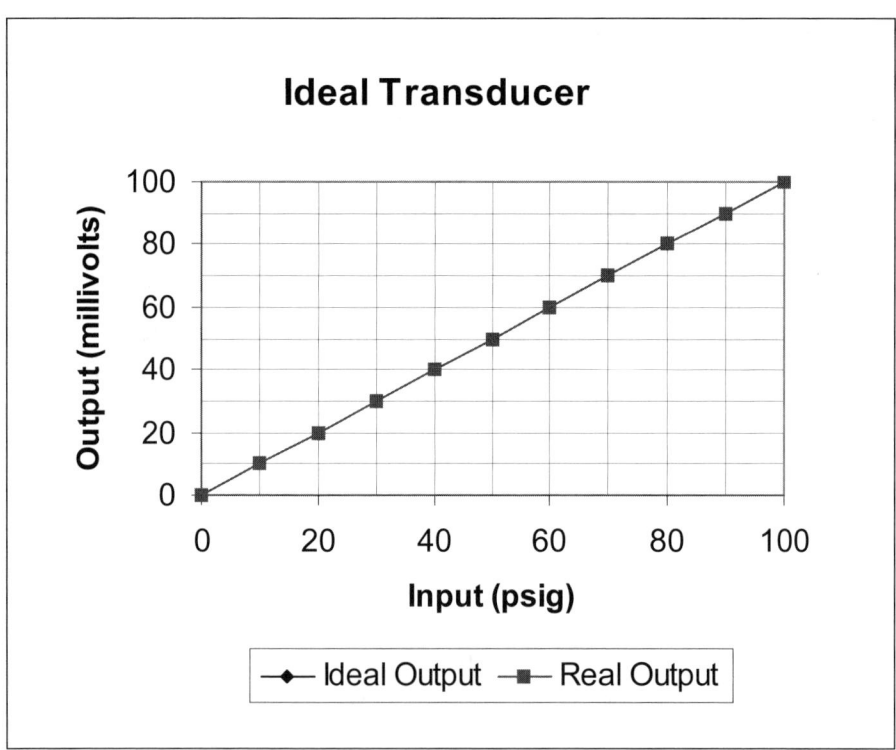

Figure 2-1. Calibration Results for an Ideal Transducer

Table 2-2 shows the results of a calibration of a real transducer in which the output readings are different from the ideal output. The errors shown in all of the tables and the resulting graphs have been exaggerated by a factor of about 100 compared to real world errors.

This has been done so the errors can be more easily seen on the graphs. The transducer's zero and span adjustments were used to set the zero output and the full-scale output so they are the same as the ideal output. This transducer did not have a linearity adjustment to minimize the errors elsewhere in the range. The maximum error of -25% of full scale that occurs at an input of 50 psig is the error specification that the transducer manufacturer would quote.

Table 2-2. Calibration Results for a Real Transducer

Input psig	Ideal Output millivolts	Real Output millivolts	Error millivolts	Error % FS %	Error % Reading %
0	0	0	0	0	
10	10	1	-9	-9	-90.0
20	20	4	-16	-16	-80.0
30	30	9	-21	-21	-70.0
40	40	16	-24	-24	-60.0
50	50	25	-25	-25	-50.0
60	60	36	-24	-24	-40.0
70	70	49	-21	-21	-30.0
80	80	64	-16	-16	-20.0
90	90	81	-9	-9	-10.0
100	100	100	0	0	0.0

Figure 2-2 shows the calibration results in a graphical form. The ideal output in this case is a straight line between the upper and lower end points of the range. The real output shows that the transducer does not produce this ideal straight line but has various errors over the range.

Figure 2-3 shows the two methods for expressing the error as previously described. The error as a percentage of full scale varies from 0 to -25% over the range. The error as a percentage of reading varies from 0 to -100% over the range.

At this point in Chapter 2, you have been introduced to the basic concepts of transducer characteristics such as zero, span, range, and error. We will now look at more specific characteristics that combine to create the total error in the transducer reading. This requires us to expand the calibration data that we previously recorded. Two or more calibra-

Fundamentals of Test Measurement Instrumentation

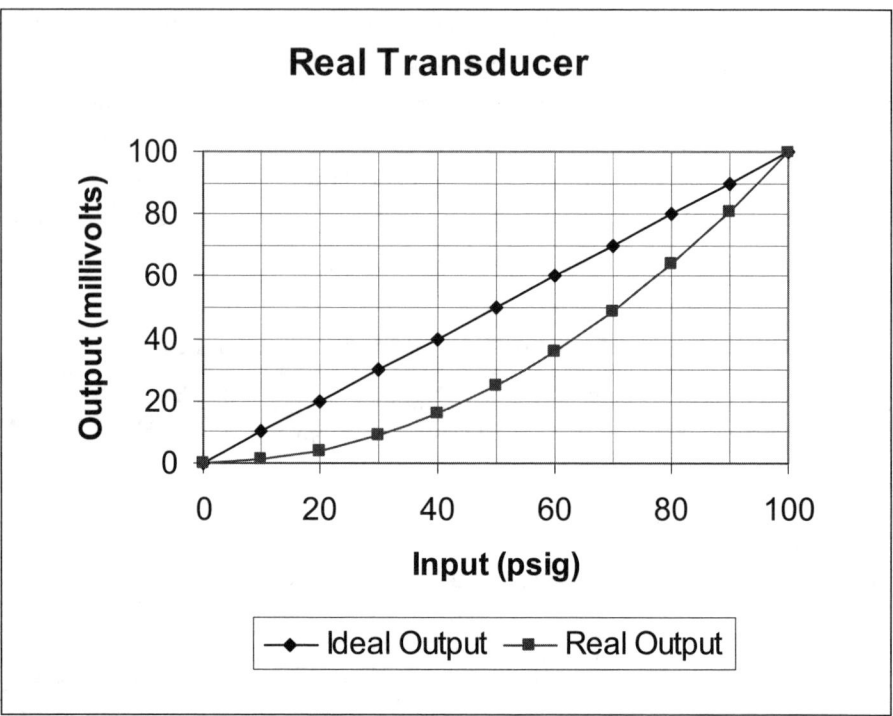

Figure 2-2. Calibration Results for an Ideal Transducer

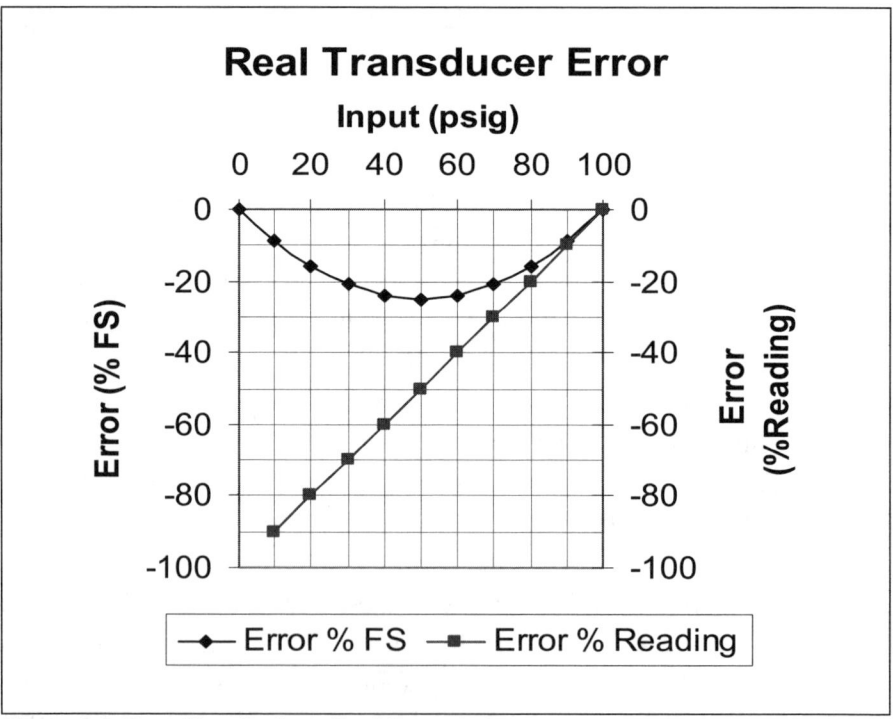

Figure 2-3. Error Results for a Real Transducer

tion cycles are needed to determine these specific characteristics. A calibration cycle involves applying known inputs over the range of the transducer, first in an ascending or increasing direction and then in a descending or decreasing direction. The input to the transducer must be applied slowly and smoothly without overshooting the input value at which an output is to be recorded. If an overshoot does occur, the input should be returned to the previous reading and then reapplied.

Table 2-3 shows the results of the expanded calibration data that was obtained by carrying out two calibration cycles. The far-right-hand Average column is obtained by averaging the data from both cycles, including the ascending and descending direction results.

Table 2-3. Results of Two Calibration Cycles

Input	Ideal Output	Increasing Run 1	Decreasing Run 1	Increasing Run 2	Decreasing Run 2	Average
psig	millivolts	millivolts	millivolts	millivolts	millivolts	millivolts
0	0	0	5	5	6	4
10	10	1	16	6	20	10.75
20	20	4	28	9	33	18.5
30	30	9	39	14	45	26.75
40	40	16	50	21	55	35.5
50	50	25	60	30	65	45
60	60	36	69	41	74	55
70	70	49	78	54	83	66
80	80	64	86	69	90	77.25
90	90	81	93	86	98	89.5
100	100	100	100	105	105	102.5

Figure 2-4 shows the results of Table 2-3 in a graphical format.

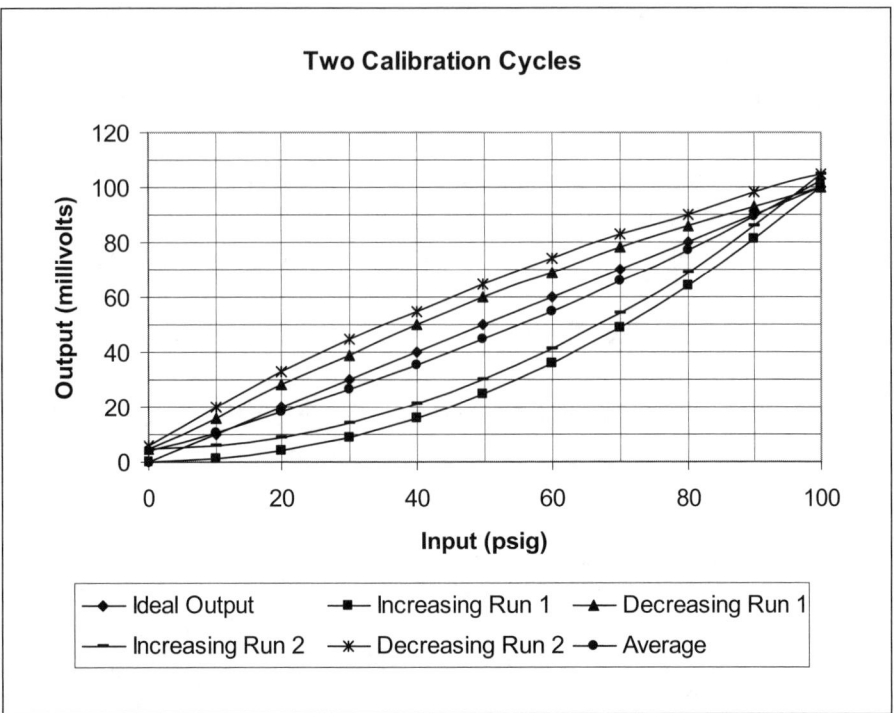

Figure 2-4. Results of Two Calibration Cycles

2.1 Hysteresis

The first static characteristic that can be determined from the calibration data is the hysteresis of the transducer. Transducer hysteresis is evidenced by a difference in output when a series of inputs is applied first in an ascending or increasing direction and then in a descending or decreasing direction. This occurs as a result of the behavior of the materials used to manufacture the sensing element or as a result of the assembly of these materials. Transducer hysteresis is most evident in sensors that use metal diaphragms, bellows, or spring assemblies in the sensing element. In a pressure transducer, the metal diaphragm is stretched when the pressure is increased and then relaxes when the pressure is reduced. Certain stress levels are built up in the diaphragm material as the pressure is increased, and when the pressure is decreased not all this stress is released. This also means that the diaphragm deflection is not the same for increasing and decreasing pressures.

The end result for transducers that measure pressure by monitoring diaphragm deflection or strain is a variation in the output signal

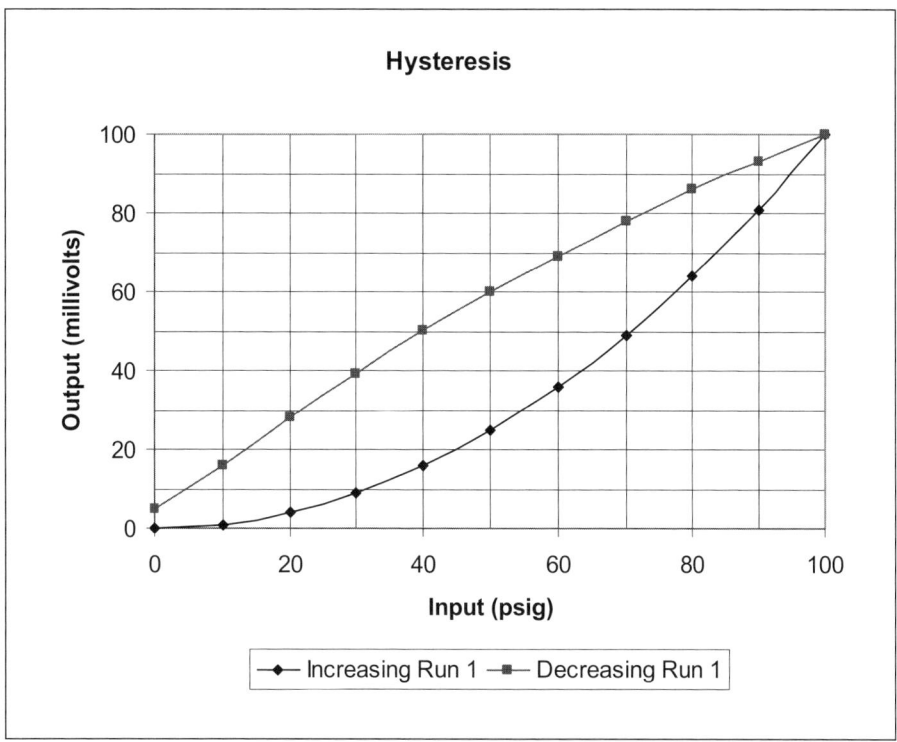

Figure 2-5. Transducer Hysteresis

depending on the direction of applied pressure. Figure 2-5 shows the results of the first calibration cycle from which we can determine the value of the hysteresis. The maximum hysteresis occurs at an input of 50 psig and is calculated by determining the difference between the output for an increasing pressure (25 millivolts) and an output for a decreasing pressure (60 millivolts). The difference is (60 − 25) = 35 millivolts, and the hysteresis is expressed as a percentage of full scale, which is (35/100) × 100 = 35% FS. Remember that the errors given in these examples have been exaggerated by a factor of 100 to highlight them graphically.

Another possible scenario that involves hysteresis is for the input to increase to some intermediate point in the range and then decrease back to the starting pressure. Table 2-4 shows the results when the input is increased to only 50 psig before it is returned to zero rather than to the full-range value of 100 psig. Figure 2-6 illustrates this partial range hysteresis. Note that the increasing values are the same as the previous example, while the decreasing values are different. This shows that hysteresis will depend on the amount of range that is covered, on the starting point, and on the previous direction in which pressure was applied. The full-range hysteresis shown in Figure 2-4 gives the worst case, and any partial range hysteresis will be smaller. The maximum hysteresis now takes place at 30 psig, and the difference between the increasing and decreasing outputs is 10 millivolts. This hysteresis is $(10/100) \times 100 = 10\%$ FS.

Table 2-4. Partial Range Hysteresis

Input psig	Increasing Partial millivolts	Decreasing Partial millivolts
0	0	2
10	1	8
20	4	14
30	9	19
40	16	23
50	25	25

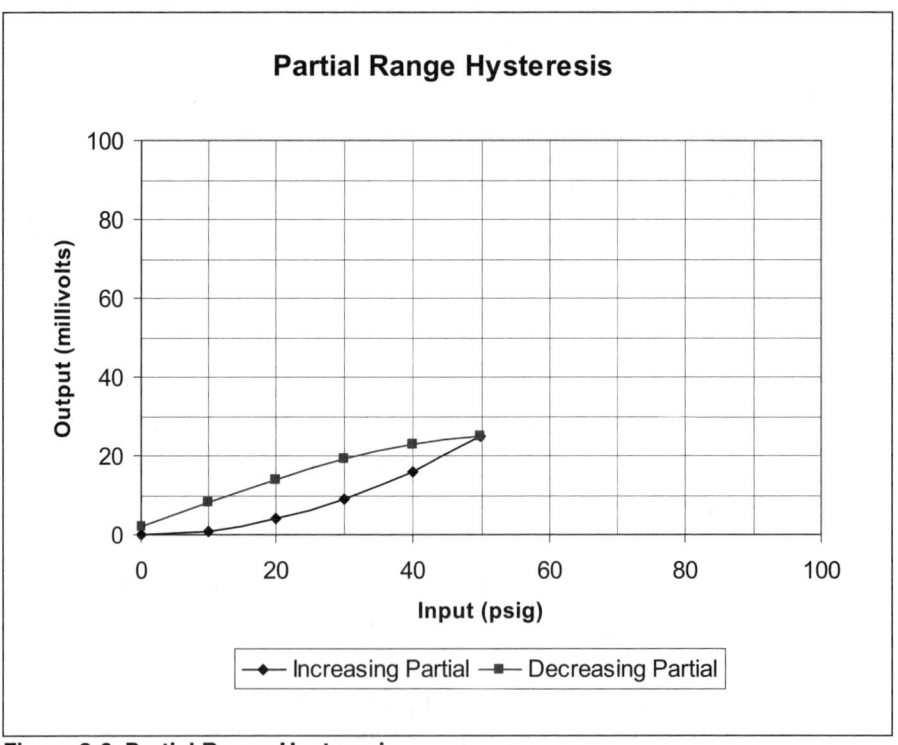

Figure 2-6. Partial Range Hysteresis

2.2 Deadband

Deadband is a transducer characteristic that is often hidden in a hysteresis curve unless specific testing is carried out to determine its presence. Deadband is the smallest change in input that causes a measurable change in the output once a change in signal direction has been made. It is primarily caused by friction in the components of the sensor or by gaps in the mating surfaces of moving components such as gears. Deadband is identified by slowly increasing the input signal in small increments until a change of output is noted. The value of input at this point is noted, and the deadband is expressed as that value expressed as a percentage of the transducer span.

Table 2-5 shows the results of a calibration performed on a transducer that has both hysteresis and deadband errors. On the increasing run, the transducer output did not change until the input was at 5 psig,

Table 2-5. Results of Calibration Performed on Transducer with Hysteresis and Deadband Errors

Input psig	Increasing Run 1 millivolts	Decreasing Run 1 millivolts
0	0.00	5.00
1	0.00	6.00
2	0.00	7.00
3	0.00	8.00
4	0.00	9.00
5	0.01	10.00
10	1.00	17.00
20	3.00	29.00
30	7.00	41.00
40	13.00	52.00
50	22.00	62.00
60	33.00	71.00
70	47.00	80.00
80	63.00	88.00
90	80.00	96.00
95	90.00	99.99
96	91.00	100.00
97	93.00	100.00
98	96.00	100.00
99	98.00	100.00
100	100.00	100.00

while on the decreasing run the output did not change until the input was at 95 psig. The deadband error in this case is stated as (5/100) × 100 = 5% of span. Once the deadband error value was exceeded, the hysteresis error starts to appear and gets added to the deadband error. Figure 2-7 shows the graphical results of a calibration of a transducer that has both hysteresis and deadband errors. Note the lack of output (the flat horizontal sections of the graph) at the start of the increasing run and this same lack of output when the direction of applied pressure is reversed at the start of the decreasing run. Unless we specifically test for the presence of deadband, it will be included in the error graph as part of the hysteresis.

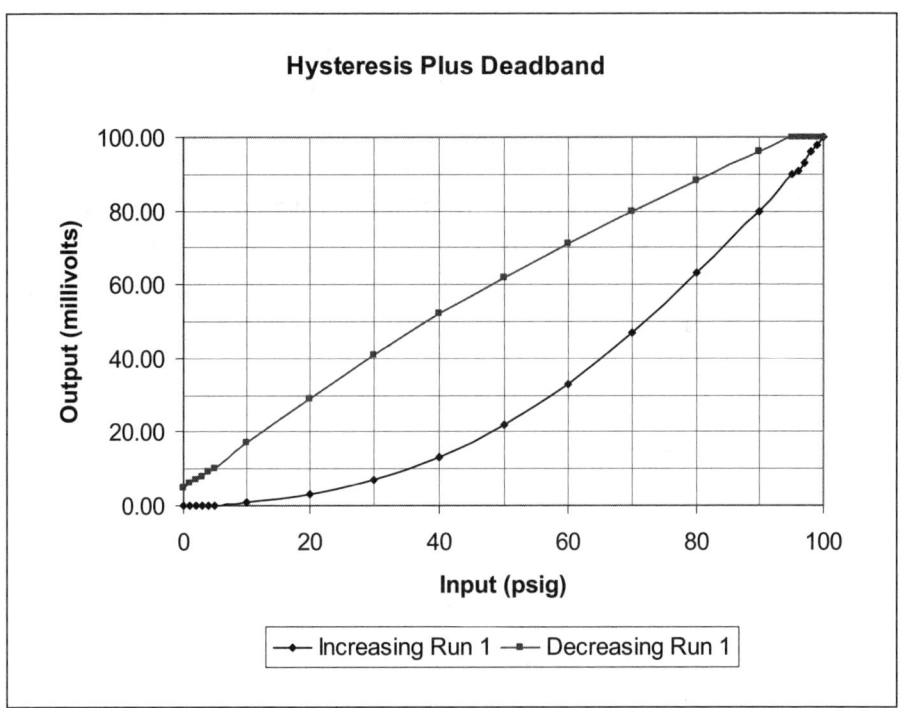

Figure 2-7. Hysteresis Plus Deadband

Table 2-6 shows the results of a calibration on a transducer that has a deadband error but does not have a hysteresis error. The output for the increasing part of the calibration cycle shows that the output does not start to change until the input exceeds 5 psig. The output then increases linearly until the upper range value of 100 psig is reached. The direction of applied pressure is then reversed and now the output does not change until the input reaches 95 psig. The output then decreases linearly until the lower range value of 0 psig is reached.

Table 2-6. Results of Calibration on Transducer with Deadband Error but Not Hysteresis Error

Input psig	Increasing Deadband millivolts	Decreasing Deadband millivolts
0	0.00	0.00
1	0.00	1.05
2	0.00	2.10
3	0.00	3.15
4	0.00	4.20
5	0.01	5.25
10	5.50	10.50
20	16.00	21.00
30	26.50	31.50
40	37.00	42.00
50	47.50	52.50
60	58.00	63.00
70	68.50	73.50
80	79.00	84.00
90	89.50	94.50
95	94.75	99.99
96	95.80	100.00
97	96.85	100.00
98	97.90	100.00
99	98.95	100.00
100	100.00	100.00

Figure 2-8 shows graphically the results of Table 2-6, a calibration of a transducer that has a deadband error but no hysteresis error.

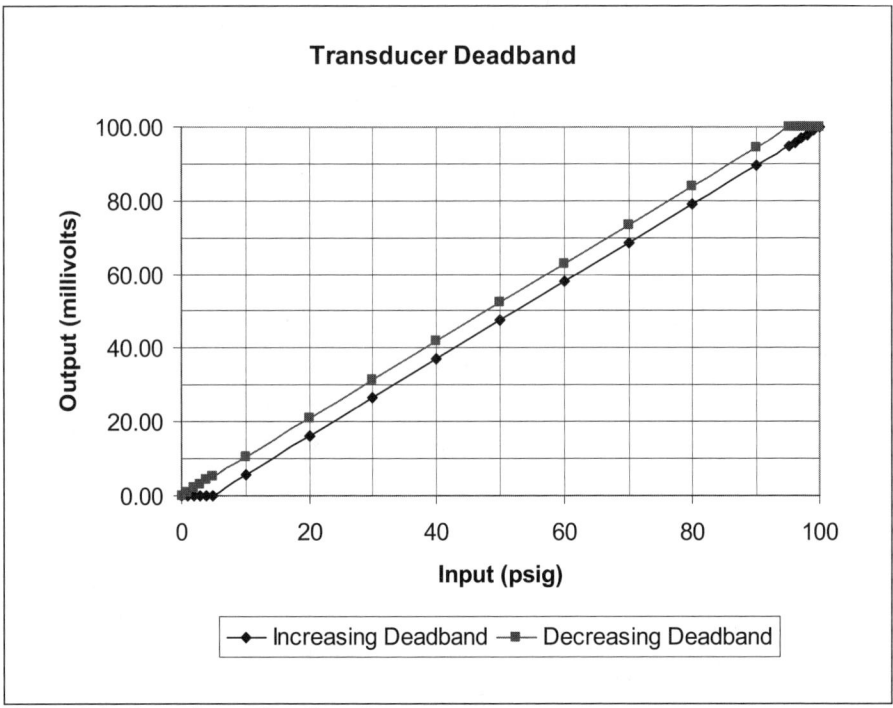

Figure 2-8. Transducer Deadband

2.3 Linearity

Linearity is a transducer characteristic that defines how closely the average calibration curve conforms to a specified straight line. The average curve is determined by averaging the results of at least two full-range input excursions in both increasing and decreasing directions. In the case of our pressure transducer, this means averaging outputs as the pressure is increased from 0 to 100 psig, decreased to 0 psig, increased again to 100 psig, and then decreased to 0 psig. The same input steps are repeated for each full-scale excursion so corresponding output values can be averaged. Averaging is performed to separate the linearity errors from the hysteresis errors. A transducer that has hysteresis (increasing and decreasing calibration data) that is symmetrical about a straight line through the calibration data will have zero linearity error with respect to that line. The averaging of the increasing and decreasing calibration data will produce a straight line.

Having a linear transducer calibration curve is an advantage since we can express the output response in terms of a simple equation, of the following form:

$$y = mx + b$$

where y represents the output in millivolts, m is the slope of the calibration line in units of millivolts per psig, x is the input in units of psig, and b is any output offset from zero in units of millivolts. The value of b can be zero, in which case the equation simplifies to:

$$y = mx$$

From this equation we can determine the output of the transducer for any given input, even if we did not take data for that input during our calibration. This equation can also be used in the software of a data acquisition system to determine the input applied to a transducer by measuring its output.

We stated earlier that linearity was defined by how closely the average calibration curve conformed to a specified straight line. This specified straight line can be fit through the calibration data in a number of different ways. The linearity is defined further depending on the method of this fit. "Independent linearity" is the error created by a line that is fit through the data in such a way as to minimize the maximum deviation of the data from the line. "Zero-based linearity" is the error created from a line that is positioned so as to coincide with the calibration curve at the lower range value and to minimize the maximum deviation throughout the rest of the range. "Terminal-based linearity" is the error created from a line that is positioned to coincide with the calibration curve at both the lower- and upper-range values. These are the most common types of specified lines. If a linearity error is not defined further it is assumed to be independent linearity.

Table 2-7 and Figure 2-9 show the averaged results compared to the ideal linear output. If there is any deviation of the averaged results compared to the ideal results then we can say that some linearity error exists.

Table 2-7. Averaged Results Compared to Ideal Linear Output

Input psig	Ideal Output millivolts	Average millivolts
0	0	4.00
10	10	10.75
20	20	18.50
30	30	26.75
40	40	35.50
50	50	45.00
60	60	55.00
70	70	66.00
80	80	77.25
90	90	89.50
100	100	102.50

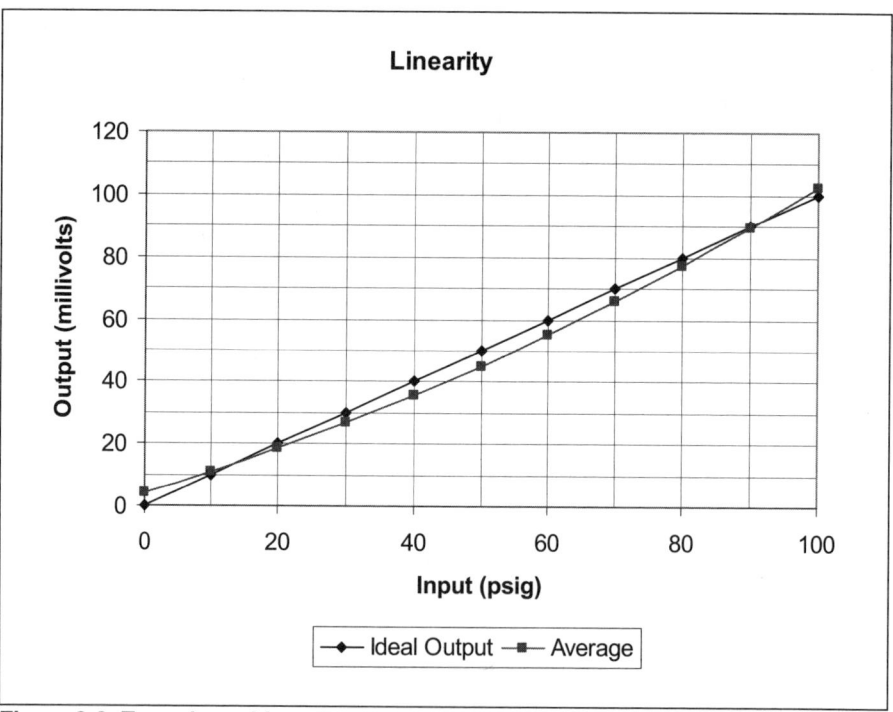

Figure 2-9. Transducer Linearity

Table 2-8 and Figure 2-10 show the results of fitting a straight line through the average curve data so as to meet the independent linearity criteria. Note that the maximum positive error is made equal to the maximum negative error. Table 2-8 shows that the error is + 4.125 millivolts at the upper- and lower-range values and –4.125 millivolts at the center of the curve. In this case, the linearity error expressed as a percentage of full-scale output is (± 4.125/100) × 100 = ± 4.125% of FS. Transducer manufacturers prefer to use independent linearity in their specifications since it provides the lowest error value.

Table 2-8. Results of Fitting Straight Line through Average Curve Data to Meet Independent Linearity Criteria

Input psig	Average millivolts	Best Fit millivolts	Linearity Error millivolts
0	4.00	-0.13	4.125
10	10.75	9.73	1.025
20	18.50	19.58	-1.075
30	26.75	29.43	-2.675
40	35.50	39.28	-3.775
50	45.00	49.13	-4.125
60	55.00	58.98	-3.975
70	66.00	68.83	-2.825
80	77.25	78.68	-1.425
90	89.50	88.53	0.975
100	102.50	98.38	4.125

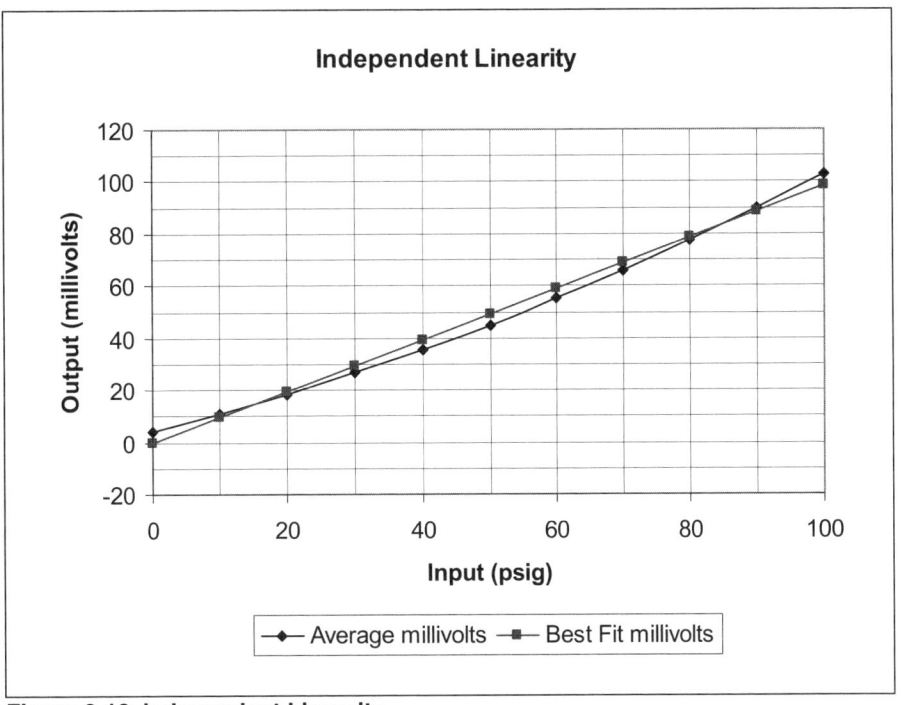

Figure 2-10. Independent Linearity

Table 2-9 and Figure 2-11 show the results of fitting a line through the average curve data so as to meet the zero-based linearity criteria. The straight line is fit through the average curve data so that the zero intercept of the line coincides with the lower-range value of the average curve. This occurs at a value of 4.00 millivolts. The slope of the line is adjusted to provide equal maximum positive and negative errors throughout the rest of the range. In this case, the maximum positive error is + 5.643 millivolts at the upper-range value and the maximum negative error of − 5.643 millivolts near the center of the range. As a percentage of full-scale output, this error is ± (5.643/100) × 100 = ± 5.643% of FS. Note that this zero-based linearity error is slightly larger than the linearity error when it is expressed as independent linearity.

Table 2-9. Results of Fitting Line through Average Curve Data to Meet Zero-based Linearity Criteria

Input psig	Average millivolts	Best Fit millivolts	Linearity Error millivolts
0	4.00	4.00	0.000
10	10.75	13.29	-2.536
20	18.50	22.57	-4.071
30	26.75	31.86	-5.107
40	35.50	41.14	-5.643
50	45.00	50.43	-5.429
60	55.00	59.71	-4.714
70	66.00	69.00	-3.000
80	77.25	78.29	-1.036
90	89.50	87.57	1.929
100	102.50	96.86	5.643

Table 2-10 and Figure 2-12 show the results of fitting a line through the average curve data so as to meet the terminal-based linearity criteria. The straight line is fit through the average curve data so the zero intercept of the line coincides with the lower-range value of the average curve. This occurs at a value of 4.00 millivolts. The other end of the line is set to coincide with the upper-range value of the average curve. This occurs at a value of 102.50 millivolts. In this case, the maximum linearity error is − 8.25 millivolts and occurs at the center of the curve. Note that this terminal-based linearity error is significantly larger than either the zero-based linearity error or the independent linearity error.

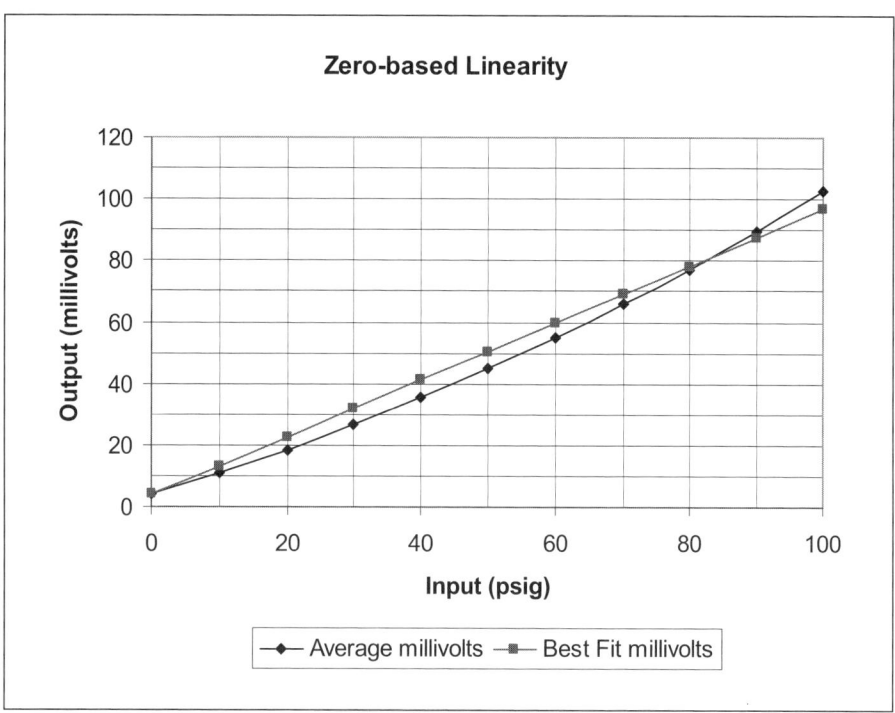

Figure 2-11. Zero Based Linearity

Table 2-10. Results of Fitting Line through Average Curve Data to Meet Terminal-based Linearity Criteria

Input psig	Average millivolts	Best Fit millivolts	Linearity Error millivolts
0	4.00	4.00	0.000
10	10.75	13.85	-3.100
20	18.50	23.70	-5.200
30	26.75	33.55	-6.800
40	35.50	43.40	-7.900
50	45.00	53.25	-8.250
60	55.00	63.10	-8.100
70	66.00	72.95	-6.950
80	77.25	82.80	-5.550
90	89.50	92.65	-3.150
100	102.50	102.50	0.000

Table 2-11 and Figure 2-13 show the results of a calibration where the average curve is s-shaped. The errors are negative for the lower half of the range and then equally positive for the upper half of the range. In this particular case, the end points of the best-fit line coincide with the

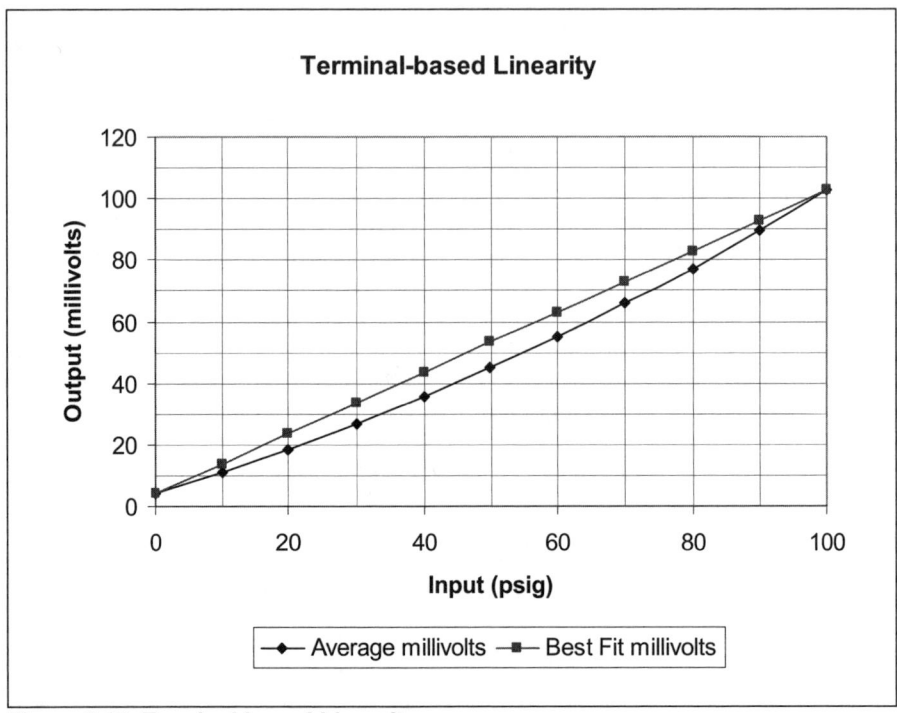

Figure 2-12. Terminal-based Linearity

upper- and lower-range values. This one best-fit line satisfies the criteria for independent, zero-based, and terminal linearity error. The maximum error value is the same for all three types of linearity.

Table 2-11. Results of Calibration Where Average Curve Is S-shaped

Input psig	Average millivolts	Best Fit millivolts	Linearity Error millivolts
0	0.00	0.00	0.000
10	7.00	10.00	-3.000
20	15.00	20.00	-5.000
30	25.00	30.00	-5.000
40	37.00	40.00	-3.000
50	50.00	50.00	0.000
60	63.00	60.00	3.000
70	75.00	70.00	5.000
80	85.00	80.00	5.000
90	93.00	90.00	3.000
100	100.00	100.00	0.000

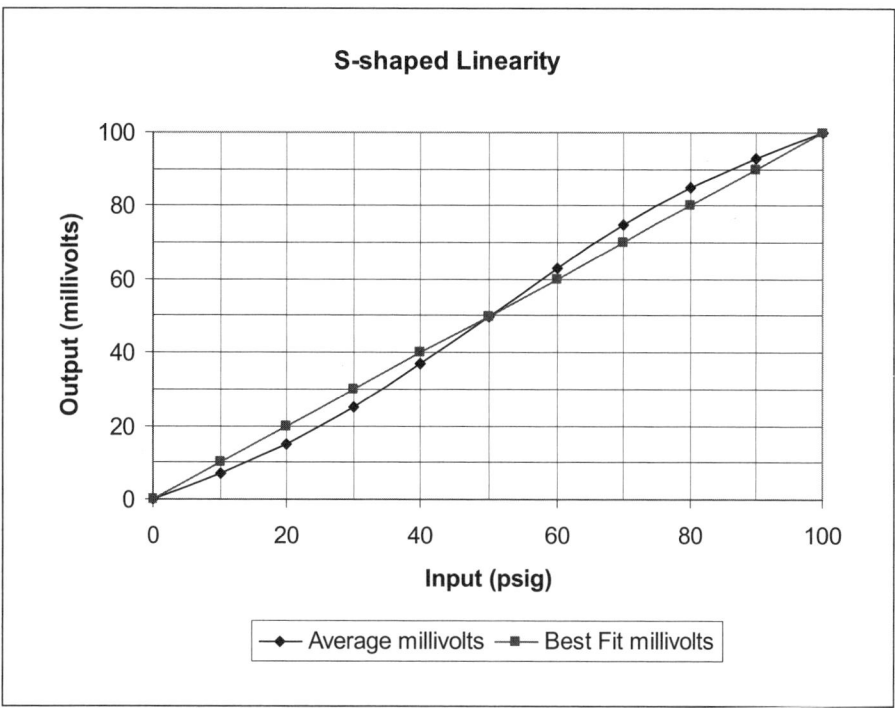

Figure 2-13. S-shaped Linearity

2.4 Conformity

Conformity is a characteristic of a transducer whose output is a specific curve rather than a straight line. Conformity error is the difference between this specified curve and the average curve created by averaging two calibration cycles. Thermocouples, resistance temperature detectors (RTDs), and thermistors are all examples of sensors that have a nonlinear output.

Conformity errors are further defined as independent conformity, zero-based conformity, and terminal-based conformity, just as linearity was. If the type of conformity is not further defined then assume it is independent conformity. Like linearity error, the conformity error is expressed as a percentage of the full-scale output (%FS).

2.5 Repeatability

Repeatability is a transducer characteristic that determines how well the transducer produces the same output when identical input values are consecutively applied under the same conditions and from the same direction. Two calibration cycles are usually used to determine repeatability. The maximum difference in output between these two calibration cycles in either the increasing or decreasing direction is the repeatability, which is expressed as a percentage of full-scale output (%FS). The fact that the two cycles must be carried out under the same conditions implies that the environmental parameters cannot change. Operating conditions such as calibration equipment, calibration procedures, and even calibration personnel should also be the same. The fact that the direction of applied input must be the same means that hysteresis is not included in the repeatability error.

Table 2-12 and Figure 2-14 show the results of carrying out two calibration cycles on a pressure transducer. The maximum output difference in either the increasing or decreasing runs is 5 millivolts. The repeatability error is then determined to be (5/100) × 100 = 5% of full scale.

Table 2-12. Results of Carrying Out Two Calibration Cycles on Pressure Transducer

Input psig	Ideal Output millivolts	Increasing Run 1 millivolts	Decreasing Run 1 millivolts	Increasing Run 2 millivolts	Decreasing Run 2 millivolts
0	0	0	5	5	6
10	10	1	16	6	20
20	20	4	28	9	33
30	30	9	39	14	45
40	40	16	50	21	55
50	50	25	60	30	65
60	60	36	69	41	74
70	70	49	78	54	83
80	80	64	86	69	90
90	90	81	93	86	98
100	100	100	100	105	105

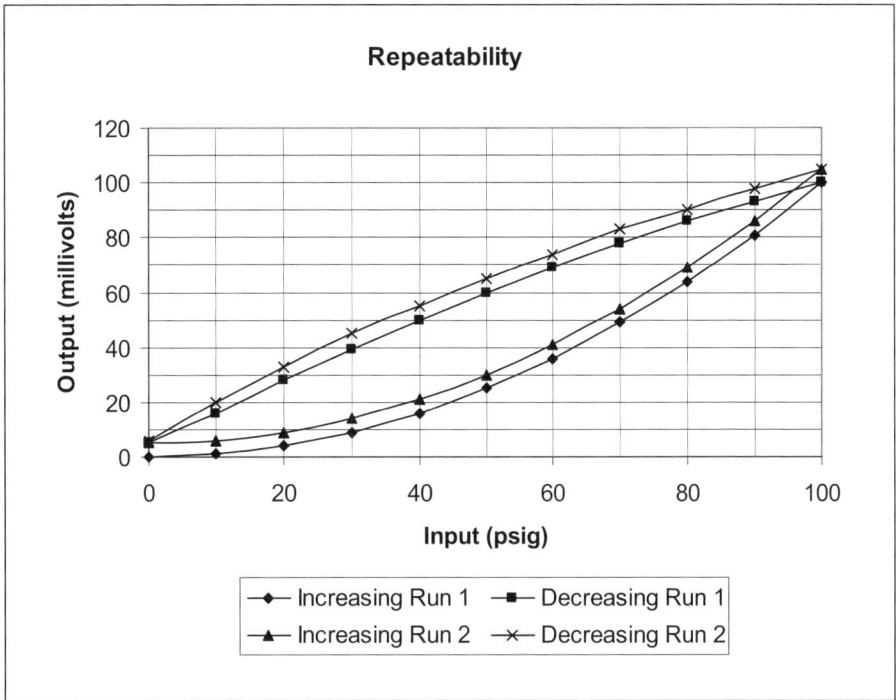

Figure 2-14. Repeatability Error

2.6 Accuracy

The errors we described in the preceding sections of this chapter will exist to some degree in all transducers. Identifying the nature of individual errors is useful to the transducer user only if it helps him or her reduce the errors by calibrating further or by accounting for the errors in the analysis of the test data. Determining the linearity error is useful if the transducer has a linearity adjustment or if the zero and span adjustments can be used to improve the fit of a line through the calibration data. Knowing the hysteresis error by itself is not of much use since the user does not usually know the direction or value of the previously occurring measurements. The term *accuracy* is used to describe the overall difference between the transducer output and the true or ideal input as determined from a calibration reference (the calibration reference has its own accuracy specification, which will be included in the determination of the transducer's overall accuracy). Accuracy includes the hysteresis, linearity, repeatability, and deadband errors.

Accuracy can be stated in a number of different ways. If the transducer or the transducer system has an output that displays the measurement in units of input, then the accuracy can be stated in those units. An example of such a system is a pressure transducer with an input range of 0 to 100 psig that is connected to a digital indicator that displays the measurement in units of psig. If the maximum positive and negative errors as determined from two calibration cycles are +3 psig and −5 psig, then the accuracy would be stated as +3, −5 psig. The accuracy can also be stated as a percentage of span, which in the case of the pressure transducer system is 100 psig. The accuracy now would be stated as +3, −5% of span. Similarly, accuracy can be stated as a percentage of upper-range value, which is 100 psig. Here the accuracy would be +3, −5% of upper-range value. A final way of stating the accuracy is as a percentage of the actual output reading. In the case of the pressure transducer system, assume that the +3 psig error occurred at an actual output of 30 psig while the −5 psig error occurred at an actual output of 50 psig. The accuracy would now be stated as ±10% of actual output reading.

2.7 Resolution

Resolution is a transducer characteristic that occurs when the sensor element produces a series of output steps in response to a continuously varying input. Examples of these types of sensor elements are wire wound potentiometers, turbine flowmeters, and digital displays. In the case of the wire wound potentiometer, a change in output is only produced when the wiper moves from one wire in the coil to the next adjacent wire. Any movement of the wiper between adjacent wires will not change the output. The turbine flowmeter consists of a rotor with multiple vanes that spins at a speed proportional to the flow. Each vane passes by a sensor, which responds to the presence of the vane and produces a pulse output. The resolution in this case is ±1 pulse. The digital display contains an analog-to-digital converter circuit that converts the continuously varying analog voltage into a proportional number of digital bits. Each additional bit produced by the analog-to-digital converter will increase the right-hand (least-significant) digit in the digital display by one count. The resolution for a digital display is ±1 digit. The transducer's resolution limits the accuracy since the accuracy specification cannot be any better than the resolution. Increasing the resolution of a system display, on the other hand, does not necessarily increase the system's accuracy. The fact that a digital display has three or four digits to the right of the decimal point rather than two digits does not necessarily mean that the system is more accurate. The information in the third or fourth digit may be less than the error in the measurement at that point. In this case, the extra resolution does not add to the accuracy of the measurement.

2.8 Environmental Effects on Transducer Characteristics

Both the transducer characteristics described so far and the characteristics we will describe in chapter 3 can be affected by any number of differences between the environment where the transducer is installed and the environment where it was calibrated. The transducer's characteristics are normally determined by testing in a benign environment such as a calibration or test laboratory, which is also referred to as "room conditions." However, the transducer may be used in a much more rugged environment where high or extreme levels of conditions such as temperature, pressure, moisture, vibration, and strain are present. At some value, these conditions will damage the transducer. Similarly, lower values will also cause permanent changes in the transducer's characteristics.

The manufacturer must therefore test the transducer under extreme conditions to determine the limits under which it can operate without being damaged or permanently affected. These limits will be stated in the transducer's specification sheet. Even lower values will cause temporary changes to the transducer's characteristics, which will disappear once the environmental condition is returned to the room conditions.

The specification sheet may quantify how much the transducer's characteristics will change in accordance with a given amount of change in each type of environmental condition. These specifications are known as "environmental sensitivity shifts." An example is the amount that the zero will shift as a result of the increase in the ambient temperature. This can be expressed as a percentage of full-scale output (%FS) per degree Fahrenheit. Knowing the sensitivity shifts and the conditions under which a measurement was made allow us to correct the test data for the effect the environment has had on the transducer.

3
Dynamic Transducer Characteristics

Chapter 2 described the transducer characteristics that are determined while the transducer is measuring a non–time varying or static signal. One of the most important characteristics of a transducer is its ability to provide an accurate measurement even when the input is changing rapidly. During rifle testing, for example, we may want to measure the pressure exerted by the explosive charge so we can calculate stresses on the rifle barrel. The pressure transducer must be capable of responding quickly to the sudden increase in pressure and providing an accurate measurement of the pressure change. Similarly, while conducting crash tests of automobiles we may want to measure the motion of crash-test dummies as the automobile is run into a solid object. The velocity of the automobile changes instantaneously from its initial speed of 60 mph to a speed of 0 mph as it impacts the solid object. The car is designed so that components and materials crumple on impact so energy is absorbed and the rate of change of velocity (acceleration) is reduced. Air bags are deployed at a certain deceleration value in order to protect the occupants. The transducers that are used to measure these changes in velocity, acceleration, and motion must be able to accurately measure changes that occur within milliseconds of the time of impact.

Each transducer system has some limit on the rate of change in input to which it can respond. This limit is determined by the transducer's design, materials used, and the capabilities of the signal conditioning equipment. The transducer manufacturer tests the transducer and states the dynamic response capabilities in the specifications.

Two examples of temperature sensors that have different dynamic response capabilities are thermocouples and resistance temperature detectors (RTDs). Thermocouples can respond very quickly to changes in temperature, while RTDs respond relatively slowly. This is due to the mass of the material that makes up the sensing element. Thermocouples can be made of extremely small-diameter wire. The only thing required to construct a thermocouple is to join the two dissimilar metal wires to form a junction. However, this junction can be a small bead that is not much bigger than the wire diameter. RTDs require that a coil of wire be formed onto a support that holds up the wire so no strain is imposed on it as temperature changes. The coil contains a relatively long wire whose entire length must change temperature in order to produce a measurable change in resistance. This takes a much longer time than it takes to change the temperature of the small thermocouple bead.

The thermocouple's design can also influence its dynamic response. If the junction can be exposed to the environment, then the response can be extremely fast. If the junction must be sheathed in a stainless steel tube to be protected against the environment, then the response is slower since the sheath as well as the junction must be heated to the new temperature. The junction in a sheathed thermocouple can either make contact with the inner wall of the sheath (grounded thermocouple), or it can be insulated from the wall (ungrounded thermocouple). The response of the grounded thermocouple is faster than the ungrounded thermocouple since the heat transfer from sheath to junction is faster than the heat transfer of sheath to insulating material (poor heat transfer rate) to junction.

Most transducers are capable of measuring both static conditions and dynamic conditions up to some specified speed limit. Measuring static conditions means that as the input is increased to a new value, the output will increase to a level that is proportional to that input, and then the output will remain at that level until the input level is changed. Such transducers are said to have "DC" or "static response capability." Some transducers will only produce an output when the input is changing faster than some specified limit. If the input stops changing, then the output will drop to zero. Such transducers are called "dynamic transducers" and are said to have an "AC response capability." Piezoelectric accelerometers and piezoelectric pressure transducers are examples of this type of transducer.

A measurement system's overall dynamic response will also be affected by the dynamic response of the signal conditioning equipment and the display equipment. Signal conditioning amplifiers will have an upper frequency response limit. This limit should be at least as high as the upper limit of the transducer if the full dynamic range of the transducer is to be used for a given test. Some amplifiers have filters built into them that will allow the user to limit the upper frequency limit to a value that matches the frequency content of the test conditions. Filtering out high-level signals that have no meaning to the test results may simplify the data analysis or simplify the display device you choose to use.

The charge amplifiers used with piezoelectric-type sensors will also have a filter to limit the low frequency limit of the signal. This is done so as to remove any erroneous data that is produced at frequencies lower than the piezoelectric transducer's low frequency limit. Displays must be selected based on the frequency of the transducer and amplifier and the frequency of the test data. Analog or digital displays are only useful for static or very slowly changing signals. Recorders are used for intermediate frequencies, while high-frequency signals require the use of oscilloscopes, data acquisition systems, and data analyzers.

The test measurement professional is responsible for selecting the system components that have the dynamic response his or her application requires. This requires that he or she know the capabilities of the system components as well as the frequencies of interest for a given test.

3.1 Frequency Response

The frequency response is the first means by which the dynamic response capability of a transducer or transducer system can be defined. To determine the frequency response, a range of sinusoidal input signals are applied to the input of the transducer. The output is monitored to determine how well it responds to the input. The frequency of the input signal is increased from a very low frequency to higher frequencies until a frequency is reached at which the transducer no longer responds to the input. These changes in frequency can either be performed in a series of steps or in a continuous sweep of frequencies. The series-of-steps method is used when the response test is carried out manually, while the frequency sweep is used when the test is carried out using a frequency response analyzer. The amplitude of the sinusoidal input signal is set to be approximately 10% of the transducer range and is centered at the midpoint of the range. If we were testing a pressure transducer with a range of 0 to 100 psig then the amplitude of the input signal would be 10 psig, and it would be centered at 50 psig. This input amplitude is maintained throughout the range of test frequencies, and the test determines how well the transducer is able to maintain the output amplitude. At some higher frequency the output amplitude will start to decrease and will continue decreasing as the frequency is increased. As the frequency is increased beyond some value, the output of the transducer will also start to lag in time behind the applied input. This lag is called the "phase shift" between the input sine wave and the output sine wave.

A reference transducer is used to monitor the input to the transducer being tested. This reference transducer must have a frequency response that is ideal over the frequency range to be tested. This means that there is no amplitude loss at higher frequencies and no phase shift. The amplitude of the output sine wave from the test transducer is compared to the amplitude of the reference transducer at a series of frequencies. The ratio of the test amplitude to the amplitude of the reference transducer is calculated and then expressed in decibels by taking the logarithm to the base ten of the ratio and multiplying the result by 20. This value is known as the "magnitude ratio":

$$\text{Magnitude Ratio (db)} = 20 \log_{10} (A_{test}/A_{reference})$$

At low frequencies, the test transducer's frequency response will be equal to that of the reference transducer. Up to some frequency, $A_{test} = A_{reference}$ and the magnitude ratio will be zero. At some higher frequency, the test amplitude will become smaller than the reference amplitude. The magnitude ratio will now become negative and will continue becoming more negative as the frequency is increased. The point at which the magnitude ratio becomes -3 db is called the "corner frequency." This is reached when the test amplitude is 0.707 times the reference amplitude.

Transducer manufacturers use this corner frequency to define the transducer's frequency response capability. A typical specification will state that the frequency response is flat from 0 Hz to the corner frequency. This is something of an exaggeration since the amplitude at the corner frequency is only 70.7% of what it should be. The phase shift described earlier will also start to occur at frequencies lower than the corner frequency. These two conditions should alert the user to the fact that the transducer should not be used to measure frequencies close to the corner frequency.

A standard (ISA-26-1968, "Dynamic Response Testing of Process Control Instrumentation") published by ISA describes the method for performing a frequency response test on process instruments. The concepts covered in that standard can be applied to test measurement instrumentation.

3.2 Response Time

The second means for defining the transducer's dynamic response is its response time. This is the time it takes for the transducer's output to reach a specified percentage of its final value when a step change is made to its input. A step change occurs when the input changes very quickly from a low value to some higher value or from a high value to some lower value. If we determine that the time it takes for the output to reach 95% of its final value is 0.1 seconds, then we say that the 95% response time is 0.1 seconds. If we determine that the time it takes for the output to reach 98% of its final value is 0.2 seconds, then we say that the 98% response time is 0.2 seconds. There is no standard percentage for defining response time, so we must state the percentage along with the time.

A similar specification for response time is the rise time. Rise time is the time it takes for the output to change from a small percentage of the final value to a large percentage of the final value. As with response-time testing, the input change required to test rise time is a step change. Unless otherwise stated, the rise time is measured from the time the output reaches 10% of the final value to the time it reaches 90% of the final value.

Another similar specification is the time constant. In a first-order system this is the time it takes for the output to reach 63.2% of the final value when a step change is made to the input. A first-order system is one that responds in the same way as a single capacitor that is being charged with current through a single resistor. The time-constant specification is useful since it can be used to determine the corner frequency in the frequency response characteristic:

$$\text{Time constant} = 1 / (2 \times \text{PI} \times f_c)$$

where:

- Time constant has units of seconds
- PI = 3.1416
- f_c = corner frequency in units of Hertz or cycles per second

3.3 Damping

Damping is the transducer characteristic that defines both how energy from a rapid change in input is dissipated within the transducer system and how it affects the transducer's dynamic response characteristics. The quick change can be the same step change in input that was used to determine the response time. A transducer is said to be "underdamped" if the output exceeds the final output (called "overshoot"), the output then oscillates above and below the final value, and the oscillations then decrease in amplitude until the output becomes steady at the final value. A transducer is said to be "critically damped" if the output does not exceed the final value and the time it takes to reach the final value is a minimum, with no occurrence of overshoot. An overdamped transducer has an output with no overshoot, but the time it takes to reach the final value is greater than that of the critically damped transducer. An increase in damping in a transducer system causes the response time and the upper limit of the frequency response to fall.

The transducer manufacturer may add damping to a transducer in order to eliminate overshoot and oscillation (called "ringing"). This can be done by electrical or electronic means or by immersing the transducer's moving components in viscous oil that slows down the speed at which the components can move.

4

Transducer Types

Chapter 4 introduces you to the types of transducers used to measure temperature, pressure, flow, displacement, velocity, acceleration, force, and load. These are the measurements that are most commonly carried out in test laboratories and production facilities.

Each of the specific measurements can be made with a variety of transducers that have different operating and physical characteristics. Descriptions of the various transducers are provided to help the user select the best transducer for a particular application.

4.1 Temperature Transducers

The transducers used to measure temperature include RTDs, thermocouples, thermistors, integrated circuit sensors, and infrared sensors. The operating principle, advantages, and disadvantages are provided for each type of transducer.

4.1.1 Temperature and Temperature Measurement

In contrast to other physical parameters, there is no common, universally accepted definition of *temperature*. Definitions for *temperature* can range from an object's relative sense of warmth or coolness to a measure reflecting the average kinetic energy associated with the motion of the molecules that make up a material. Temperature is an intrinsic property of matter that quantifies the ability of one body to transfer heat energy to another body. Heat flows from a high temperature area or body to a low temperature area or body. If a high-temperature body is put in contact with a low-temperature body, heat energy will decrease in the hot body and increase in the cooler body until a thermal equilibrium is reached. At the molecular level, the velocity of the molecules in the hot body will decrease as they collide with the molecules in the cooler body. The velocity of the molecules in the cooler body will increase as they are hit by the molecules in the hot body. At equilibrium, the velocities are the same in both bodies, and we can say that the temperatures of the two bodies are the same. If heat energy is added to a body, then the velocity of the molecules increases and the temperature of the body increases.

We have an intuitive sense of temperature since, over a fairly limited range, we can feel by touch an object's degree of hotness or coldness. In this case, our skin, nerves, and brain form the temperature transducer. Temperature transducers, as described in this section, have been devised to measure temperature over an extremely wide range, from absolute zero to thousands of degrees.

Early versions of temperature transducers were based on the expansion in a liquid's volume that occurs when the liquid's temperature is increased. These thermometers were constructed by connecting a liquid-filled bulb to a capillary tube. The expansion of the liquid, when heated, caused the liquid to rise in the tube. A linear scale on the side of the tube was used to measure the height of the liquid, and this height was proportional to the temperature. This type of transducer evolved into the liquid-filled glass thermometer used in homes, hospitals, kitchens, and classrooms. To make the liquid height useful, some readily reproducible reference points had to be used to denote points on the scale. Numbers were assigned to these points, a number of graduations between reference points were assigned, and a temperature scale was thus created.

Gabriel Fahrenheit created such a temperature scale in 1724 using a mercury-in-glass thermometer. He assigned zero degrees to the temperature of a mixture of salt, water, and ice. A second point was the temperature of a mixture of water and ice; he originally assigned a value of 30° to this temperature. A third point was the body temperature of a healthy man, which was assigned a temperature of 96°. Given this thermometer scale, the temperature of boiling water was determined to be 212°, and Fahrenheit then changed the 30° value to 32° so the difference between the freezing point and the boiling point was 180°. This scale became the Fahrenheit temperature scale.

Later, Anders Celsius created a scale that assigned 100° to the temperature of a mixture of ice and water and 0° to the boiling point of water. One hundred graduations, called a centigrade, were used to divide the scale into 100°. The 0° and 100° were later reversed to create the familiar centigrade temperature scale, where 0° is the temperature of a mixture of ice and water and 100° is the temperature of boiling water.

In the late 1880s, experiments with gases led to the finding that gases at low pressures behave like an ideal gas and have the following relationship between pressure, volume, and temperature:

$$pV = (constant)\, t$$

From this equation we can see that, for a constant volume, there is a linear relationship between temperature and pressure. We can also see that there is a defined zero temperature that occurs when the gas pressure is zero. This temperature is called "absolute zero." Using this zero point and the triple point of water as a reference, the absolute temperature scale was created with units of Kelvins (K). The triple point of water, which occurs when the solid, liquid, and vapor phases of pure water are in equilibrium, was assigned the value of 273.16 K.

In 1948 the centigrade scale was replaced by the Celsius scale in which the triple point of water was assigned the value of 0.01°C, and a degree on the Celsius scale represented the same temperature change as a degree on the absolute temperature scale.

The relationships between these three temperature scales are as follows:

$$°C = (°F - 32) \times 5/9$$

$$°F = (°C \times 9/5) + 32$$

$$K = °C + 273.15$$

4.1.2 Thermocouples

Thermocouples are probably the most widely used type of temperature transducer (see Figure 4-1). They are constructed by joining two wires made of different materials and placing the junction at the point at which we want to measure the temperature. A small voltage will appear at the ends of the wires opposite the junction if there is a difference in temperature between the junction location and the measuring location. The magnitude of the voltage is proportional to the temperature difference and will vary depending on the types of material used in the two wires. The relationship between voltage and temperature is nonlinear, and the curves are different for each pair of wire material used.

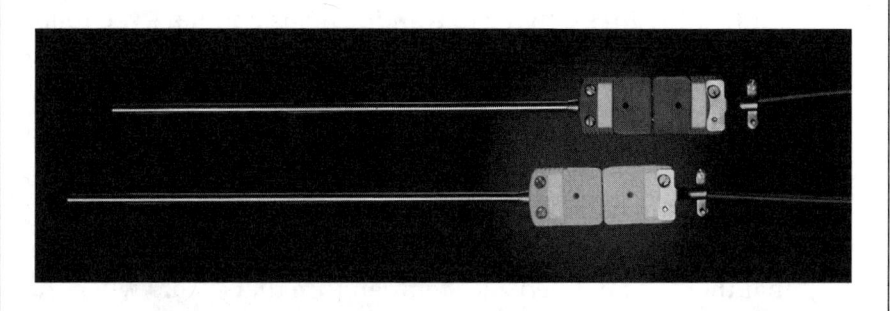

Figure 4-1. Stainless Steel Sheathed Thermocouples
(© Copyright Omega Engineering, Inc. All rights reserved. Reproduced with the permission of Omega Engineering, Inc., Stamford, CT 06907. www.omega.com)

The electromotive force (emf) is generated by the wires, not by the junction, and the emf is present whenever a temperature gradient exists along the wire. Two different materials will produce different emfs for the same temperature gradient. It is this difference that we measure. Tables of emfs versus temperature are available from manufacturers or standards organizations (see, for example, ISA-MC96.1-1982, "Temperature Measurement Thermocouples").

A number of standard wire pairs are normally used, and each has been given a name, a letter designation, and a color code that is used on the wire insulation. The name given each wire pair is based on the combination of the two different wire materials. The one that produces the more positive emf is listed first in the name. The wire insulation color for the negative wire is red, while the positive wire insulation color depends on the combination of wires used.

Four wire combinations commonly used for relatively moderate temperatures are as follows:

Type T
Name:	copper-constantan
Positive wire:	copper
Positive wire insulation color:	blue
Negative wire:	constantan (copper, nickel alloy)
Negative wire insulation color:	red
Temperature range:	-454 to 752°F maximum, -328 to 662°F useful

Type K
Name:	Chromel-Alumel
Positive wire:	Chromel (chromium, nickel alloy)
Positive wire insulation color:	yellow
Negative wire:	Alumel (aluminum, nickel alloy)
Negative wire insulation color:	red
Temperature range:	-454 to 2501°F maximum, -328 to 2282°F useful

Type J
Name:	iron-constantan
Positive wire:	iron
Positive wire insulation color:	white
Negative wire:	constantan (copper, nickel alloy)
Negative wire insulation color:	red
Temperature range:	-346 to 2193°F maximum, 32 to 1382°F useful

Type E
Name:	Chromel-constantan
Positive wire:	Chromel (chromium, nickel alloy)
Positive wire insulation color:	purple

Negative wire: constantan (copper, nickel alloy)
Negative wire insulation color: red
Temperature range: -454 to 1832°F maximum, -328 to 1652°F useful

The maximum useful temperature is also limited by the wire size and insulation materials used to construct the thermocouple. The color codes just presented are commonly used in North America; different color codes are used internationally.

Thermocouple wire is available with three different calibration grades called standard grade, premium or special grade, and extension grade. The error of the premium grade wire is approximately one half that of the standard grade. Extension-grade wire suffers from errors larger than that of the standard grade wire. Extension-grade wire is used to connect the thermocouple to a readout device located a long distance from the thermocouple. It is assumed that the difference in temperature over the length of the extension wire will be small, and so any errors created will be small compared to the overall temperature being measured. Unless a large number of thermocouples are being run over a great distance, however, it is better to use premium grade wire throughout the installation. The benefits of increased measurement accuracy will override the relatively small increase in wire cost.

Thermocouples that are constructed by making a junction between two dissimilar material wires are used mainly in research and development laboratories for short-term tests. The advantages they offer include the ability to use fine wire for fast-response measurements and to construct the thermocouple to suit the physical demands of the test equipment. Examples of these unprotected wire thermocouples are shown in Figure 4-2. Permanent installations that require a more rugged and stable thermocouple have led to the stainless steel sheathed industrial thermocouple as shown in Figure 4-3. These are constructed by inserting the wires and a powdered insulation such as magnesium oxide into a small-diameter stainless steel protective sheath. The sheath is drawn through dies to reduce the diameter and to compact the insulation around the wires. The junction can be exposed to the environment, sealed and grounded to the sheath, or sealed and not grounded to the sheath.

Chapter 4 – Transducer Types   59

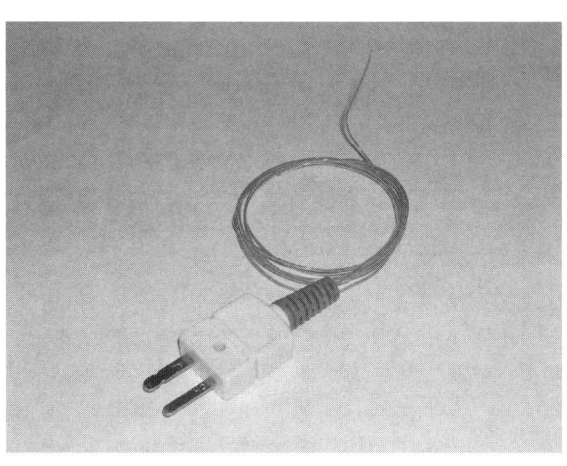

Figure 4-2. Unprotected Wire Thermocouples
(© Copyright Omega Engineering, Inc. All rights reserved. Reproduced with the permission of Omega Engineering, Inc., Stamford, CT 06907. www.omega.com)

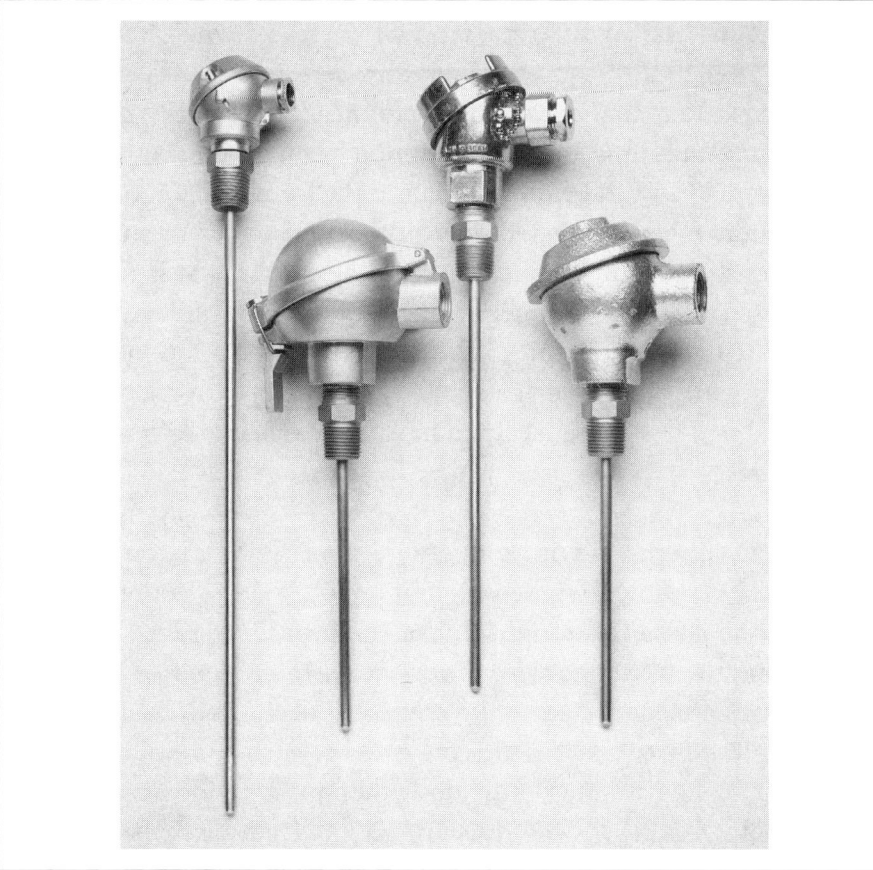

Figure 4-3. Steel Sheathed Industrial Thermocouples
(© Copyright Omega Engineering, Inc. All rights reserved. Reproduced with the permission of Omega Engineering, Inc., Stamford, CT 06907. www.omega.com)

The voltage output from a thermocouple is related to the difference in temperature between the junction end of the wires and the open end of the wires where the voltage is measured. This means that the temperature at the measuring end must be known or be kept at a known constant temperature. In most modern signal conditioning units or displays, the temperature of the terminal strip where the thermocouple wires are attached is measured. This reading is then used to adjust the display for any changes from an assumed temperature for the terminal strip. This adjustment is carried out automatically by the software or firmware used by the signal conditioning or display equipment. The use of miniature temperature-controlled ovens or chillers to maintain the terminal strips at a known constant temperature has largely been superseded by this more recently developed method.

4.1.3 Resistance Temperature Detectors

The resistance of a metal will increase when its temperature is increased. Resistance temperature detectors or RTDs are temperature transducers that apply this phenomenon in order to measure temperature. Platinum is the most common metal used to construct RTDs. Platinum RTDs can be made either in the form of a coil of fine wire or as a film of metal deposited on a substrate. The coil or substrate is usually encased in a thin-walled stainless steel tube that is filled with a powdered insulating material such as magnesium oxide. This tube is sealed at the coil end to protect the coil from the measurement environment. Wires from each end of the coil or film are brought out of the other end of the tube, and a moisture resistant seal is made around the wires.

RTDs are made with a two-wire, three-wire, or four-wire configuration. In the two-wire configuration, one wire is connected to each end of the coil and brought out of the protective tube. In the three-wire configuration, one wire is connected to one end of the coil, and two wires are connected to the other end of the coil. In the four-wire configuration, two wires are connected to each end of the coil. The different configurations are designed to provide various means of compensating for the resistance of the lead wires that are used to join the RTD to the signal conditioner or readout.

The two-wire RTD does not have any means of compensating for lead resistance and is only used when the RTD coil is connected directly to the readout device. With a two-wire configuration, the only method for measuring coil resistance is to measure the resistance at the ends of the two lead wires. This means that the total resistance of the two lead wires is added to the resistance of the coil, and the measurement is high. If the two lead wires were only 25 feet long and made of 16 gauge copper wire, the temperature measured with a 100-ohm platinum RTD would be high by approximately 1°F.

The three-wire configuration is connected to a Wheatstone bridge circuit that adds the lead resistance into opposing arms of the bridge. If the lead resistances are equal, this causes the additional resistance to be canceled, and the voltage output of the bridge is proportional to the resistance of the coil. If the resistances of the lead wires change equally due to thermal exposure, the output voltage is not affected. If the contact resistance of terminals or connectors changes due to corrosion and the change is not equal in all leads, then the bridge output will have an error. The three-wire configuration is the one most commonly used today.

In the four-wire configuration, one wire from each end of the coil is used to carry a constant current from the signal conditioner to the RTD. The resulting voltage across the RTD is measured using additional wires that are attached to each end of the coil. This voltage is proportional to the resistance of the coil. The high impedance of the voltage measuring device means that the lead wire resistance has no effect on the voltage measurement.

Platinum is the most common metal used to make RTDs since it can be produced in a pure form that allows elements to be manufactured that have a very reproducible resistance-versus-temperature relationship. This resistance-versus-temperature relationship in platinum is also highly linear compared to other metals. This characteristic allows simple circuits or software to be used to produce signal conditioning that has a linear output versus temperature. The RTD is also a very stable device as long as it is not subjected to physical impacts or heavy vibration. Other materials used for RTDs include copper, nickel, and an iron-nickel alloy called Balco.

Another characteristic of the RTD is its nominal resistance at 0°C (R_0). The most common value for platinum RTDs is 100 ohms; values of 200 ohms and 1000 ohms are used less frequently. Platinum RTDs that are used to calibrate other temperature sensors are called "standard platinum resistance thermometers" (SPRTs) and usually have a resistance of 25.5 ohms at 0°C.

Another specification that defines the RTD's resistance-versus-temperature relationship is called the alpha, alpha coefficient, or temperature coefficient. This figure is the average slope of the resistance-versus-temperature curve between 0°C and 100°C, normalized to the resistance at 0°C. The units of the coefficient are ohms per ohm per degree Celsius:

$$\text{alpha} = (R_{100} - R_0) / (R_0 \times 100 \,°C)$$

where:

R_{100} is the resistance of the RTD at 100°C
R_0 is the resistance of the RTD at 0°C

The value of alpha is defined in the standard applicable for the particular RTD. Several standards are used today, and the value of alpha is different in each one.

The standard alpha of IEC (International Engineering Consortium) and of DIN (Deutsches Institut für Normung e.V.) has a value of 0.003850 for a platinum RTD.

The U.S. industrial standard alpha has a value of 0.003902 for a platinum RTD.

The old U.S. standard alpha has a value of 0.003920 for a platinum RTD. The Japanese standard alpha has a value of 0.003916 for a platinum RTD.

These differences mean that the RTD and the signal conditioning or display must share compatible values of alpha.

An error will occur if the units are mismatched.

4.1.4 Thermistors

The name *thermistor* is a contraction of *therm*ally sensitive res*istor*. Like RTDs, thermistors are sensors that change resistance when their temperature changes. The resistance can increase with an increase in temperature (positive temperature coefficient or PTC) or decrease with an increase in temperature (negative temperature coefficient or NTC). The most commonly used NTC thermistors are semiconductor resistors made from the oxides of manganese, cobalt, copper, and nickel.

The NTC thermistor has a high resistance at room temperature (thousands of ohms) and produces a large resistance change per degree of temperature change. This high sensor resistance and large temperature coefficient mean that the errors caused by lead-wire resistance are not an issue with NTC thermistors as they are when using RTDs.

The resistance change with temperature change is highly nonlinear for thermistors compared to RTDs. The relationship is described by the Steinhart-Hart equation for NTC thermistors as follows:

$$\text{Temperature (°C)} = (1 / (a + b(\ln R) + c(\ln R)^2 + d(\ln R)^3)) - 273.15$$

The a, b, c, and d coefficients are provided by the thermistor manufacturer. They are specific to the type of thermistor, its resistance at 25°C, and the defined temperature range. Thermistors' outputs have not been standardized to any great degree in comparison to RTDs and thermocouples. This means that probes and displays are generally matched as a pair, and one manufacturer's probe cannot be substituted for the probe of another.

The basic thermistor element can take the form of a tiny glass bead, a disk, or a chip. Bead type thermistors are shown in Figure 4-4. These elements are usually too fragile to be used in these forms, however, so the element is encased in a thin-walled steel sheath for protection as shown in Figure 4-5. The element's small size means that the sensor has a very small response time compared to the RTD.

Thermistors' temperature range is typically smaller than that of the RTD and thermocouple. The typical range for thermistors is from −50°C to 200°C.

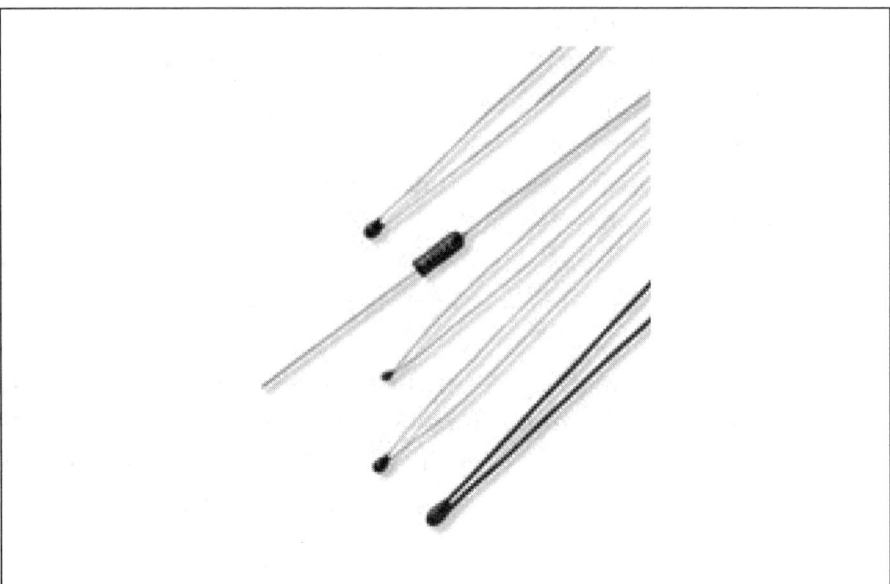

Figure 4-4. Bead Type Thermistors
(Courtesy, Quality Thermistor, Inc., www.thermistor.com)

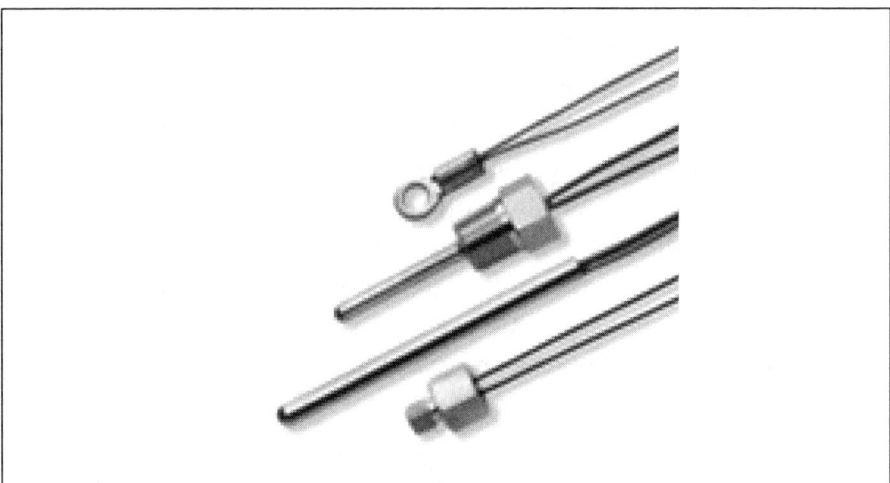

Figure 4-5. Stainless Steel Sheathed Thermistors
(Courtesy, Quality Thermistor, Inc., www.thermistor.com)

The major advantage of the thermistor is its large output change per degree of temperature change. This characteristic means that the user can detect very small changes in temperature that the RTD or thermocouple would not sense.

4.1.5 Integrated Circuit Sensors

The integrated circuit temperature transducer is based on a relationship of voltage to current of semiconductor diodes that is sensitive to temperature. This is a classic case of a disadvantage in a device being utilized to produce a practical sensor. The integrated circuit temperature transducer operates over a relatively small temperature range compared to RTDs and thermocouples. This range is normally limited to -25°C to 150°C, but for some products the range extends from -55°C to 200°C.

Integrated circuit temperature transducers are normally constructed in the typical integrated circuit packaging, such as a transistor package as shown in Figure 4-6 or an in-line circuit board mounting package. Signal conditioning circuitry can be assembled within the same package. This results in a transducer that has a high-level signal that can be easily integrated into other electronics. Transducers that have digital outputs are also available, simplifying the use of transducers in microcomputer-based devices. The mass production techniques used in semiconductor manufacturing allow these devices to be made inexpensively.

Figure 4-6. Integrated Circuit Temperature Sensor

4.1.6 Infrared Sensors

Infrared temperature sensors are different from RTDs, thermocouples, and thermistors in that they measure an object's temperature without contacting it. They do this by measuring the amount of infrared energy that is emitted from the object's surface. Infrared energy is part of the electromagnetic spectrum, which also contains other light waves such as visible light, ultraviolet light, and x-rays. The wavelength of infrared energy varies from 0.7 microns to 1000 microns. Infrared temperature transducers use the range from 0.7 to 14 microns.

The amount of energy an object emits can be described in terms of its radiance expressed in units of watts per unit area. A higher radiance level means that the temperature of the object is higher. When the radiance at all infrared wavelengths is at a maximum for a given temperature, the object is described as a "blackbody." For a blackbody, we can now determine a relationship between its radiance and its temperature. Since most materials do not emit the maximum radiance for a given temperature, we must know the fraction of radiance that the material emits compared to the radiance of a blackbody at the same temperature. This fraction or ratio is known as the "emissivity" of the surface of the material. The value of emissivity varies from 0 for a shiny mirror to 1.0 for a blackbody. Many painted, oxidized, and organic surfaces have an emissivity of 0.95, and sensors with a fixed emissivity setting often use this value. Other sensors allow the emissivity setting to be adjusted to suit the application.

Extensive lists of emissivity values are available and the value you select must be based on the material, the surface condition, and, in some cases, the temperature range. If you cannot find a value of emissivity, you must test the material to determine the temperature using a contact sensor such as a thermocouple. You then adjust the emissivity setting of the infrared sensor until the temperature output agrees with the thermocouple reading. In some cases, you can coat the material's surface with a flat black paint and use an emissivity of 0.95 for the measurement.

The advantages of the infrared sensor include its high upper-range limit of approximately 3000°C and its ability to measure temperature without contacting the test object. Its disadvantages include its rela-

Table 4-1. Comparison of Temperature Sensors

Sensor Type	Advantages	Disadvantages
Thermocouples	inexpensive rugged fast response standardized output wide measurement range no self-heating	moderate accuracy subject to drift or instability nonlinear low signal output cold junction reference required
RTDs	good accuracy good stability standardized output good linearity low self-heating	slow response sensitive to shock and vibration large sensing element lead-wire sensitive
Thermistors	fast response high sensitivity small size lead-wire insensitive	nonstandard output highly nonlinear poor stability high self-heating
Integrated circuit sensors	inexpensive linear output small size high sensitivity	narrow temperature range moderate accuracy slow response
Infrared sensors	no target contact wide temperature range fast response good stability	emissivity sensitive large size high cost environment sensitive

tively high cost, large size, and the requirement to establish the emissivity. You should always remember that you are making a surface temperature measurement with an infrared sensor. This surface temperature may not be the same as the bulk temperature of the test object.

4.2 Pressure Transducers

There is an extremely large number of pressure transducers available in today's test instrumentation marketplace. These pressure transducers use several different operating principles and are designed to operate in a wide variety of test conditions. The types of pressure transducers include strain gauge, piezoelectric, piezoresistive, capacitive, variable reluctance, resistive, and integrated circuit. Table 4-2 provides a summary of the advantages and disadvantages of the various types of pressure transducers.

Table 4-2. Comparison of Pressure Sensors

Sensor Type	Advantages	Disadvantages
Strain Gauge	Capable of DC response	Large size Limited temperature range Low sensitivity
Piezoelectric	Small size Wide temperature range Fast response Self-generating signal Rugged	AC response only Vibration sensitive Low sensitivity High output impedance Temperature sensitive
Piezo-resistive	Small size DC response High sensitivity Fast response	Temperature sensitive
Capacitive	Low hysteresis Good linearity Good stability	Complicated signal conditioning Large size
Variable reluctance	High sensitivity Rugged Fast Response	Large size
Resistive	Low cost	Slow response Subject to wear High hysteresis Low repeatability Large size
Integrated circuit	Low cost Small size	Limited temperature range Limited pressure range

4.2.1 Pressure and Pressure Measurement

Pressure is defined as a force per unit area:

$$P = F/A$$

Pressure measurements can be characterized as being either absolute or relative. Absolute pressures are those pressures above zero, which is the pressure that occurs in a perfect vacuum. Relative pressures are those pressures above or below the atmospheric pressure. Atmospheric pressure is the pressure that is exerted by the height (and thus weight) of the air above the measurement point. Pressures above the atmospheric pressure are called "gauge pressure" while those below atmospheric pressure are called "vacuum pressure."

In transducers and other pressure measurement instruments that indicate gauge pressure or vacuum pressure, the pressure being measured acts on one side of the sensing element (diaphragm, bellows, or bourdon tube), while the atmospheric pressure acts on the other side. The sensing element and thus the transducer output respond to the difference between the measured pressure and the atmospheric pressure. As we all know from weather reports, the atmospheric pressure can vary from hour to hour and day to day. The atmospheric pressure will also decrease as we go higher in altitude, where the height of the air above the measurement point gradually diminishes. These variations are relatively small and usually not significant unless the measurement being made is close to atmospheric pressure.

It should always be remembered, however, that the gauge and vacuum measurements are being made relative to the atmospheric pressure, which is a variable reference. In transducers that indicate absolute pressure the pressure being measured acts on one side of the sensing element while the other side is connected to a chamber that has been evacuated of air to be as close to a perfect vacuum as possible. Variations in atmospheric pressure or altitude will not affect the output of these transducers when they are used in a system that is sealed from the atmosphere.

The relationships between the various pressures are as follows:

absolute pressure = gauge pressure + atmospheric pressure

or:

gauge pressure = absolute pressure − atmospheric pressure

Examples of Absolute and Gauge Pressure Measurements

Example 1: We inflate the tire of a car to 30.0 psi (gauge) using a tire gauge that has a bourdon tube pressure gauge as its indicator. We find from the local weather station that the atmospheric pressure is 14.7 psi. An absolute pressure transducer connected to the tire will show a pressure of 47.7 psi (absolute).

We drive the car up a mountain where the atmospheric pressure is 14.5 psi due to the increased altitude. The absolute pressure transducer will still show a pressure of 47.7 psi (absolute) since the same number of gas molecules is still impacting the interior of the tire wall, exerting the same pressure. The tire gauge will now show a tire pressure of 30.2

psig. To maintain the original tire stiffness we should reduce the tire pressure to 30.0 psi (gauge) as measured by the tire gauge. In this example, it is correct to use a gauge pressure measurement since the tire stiffness is determined by the difference between the pressure acting on the inside of the tire and the pressure acting on the outside of the tire (the atmospheric pressure). This difference is the pressure that a gauge pressure measurement indicates.

Example 2: We want to determine the height of water in a tank that has its top open to the atmosphere. We can do this by mounting a pressure transducer in a penetration through the wall of the tank near the bottom. The pressure exerted on the front of the transducer diaphragm will be the sum of the pressure caused by the height of the water above the transducer and the atmospheric pressure acting on the surface of the water. In the case of a gauge pressure transducer, the atmospheric pressure will also act on the back side of the transducer diaphragm. This pressure will cancel out the atmospheric pressure acting on the front of the transducer, and the gauge pressure transducer will indicate the pressure caused by the height of water alone. An absolute pressure transducer will indicate the sum of the atmospheric pressure plus the pressure caused by the height of water.

We could adjust the zero of the transducer readout to remove the pressure caused by atmosphere, thus leaving only the pressure caused by the height of water. This adjustment would only be true for one given atmospheric pressure, and the indication of the absolute pressure transducer will be in error for any other atmospheric pressure. In this example, it is correct to choose a gauge pressure transducer since the changes in atmospheric pressure act on both the front and back of the transducer diaphragm. The gauge pressure transducer responds only to the pressure exerted by the height of water in the tank, which is the measurement we wish to make.

Example 3: A supplier of natural gas measures the flow of gas to its customer by using a flowmeter that indicates the volume of gas per unit time (cubic feet per second). This is called a "volumetric flowmeter." The customer will want to pay for the gas based on the energy produced when the gas is burned. This energy is proportional to the mass of the gas, not its volume. The supplier must continuously determine the gas's density (or specific gravity) by measuring the flowing pressure (and temperature). The density is determined by entering the pressure and temperature measurements into a formula in which the units of pressure and temperature are absolute units. The density (pounds per cubic foot) is then multiplied by the volumetric flow (cubic feet per second) to obtain the mass flow in units of pounds per second. The supplier and customer now know the mass of gas delivered regardless of any changes of pressure and temperature that occurred while the gas was flowing. In this example, selecting an absolute pressure transducer was the correct choice since the density determination formula requires the absolute pressure as an input.

Units of Pressure

The SI unit of pressure is the Pascal (Pa), which is a force of one Newton (N) acting on an area of one square meter (m^2). The Pascal represents a relatively small pressure, and so more commonly used units are the kilopascal (kPa) and the megapascal (MPa). A common U.S. unit of pressure is the psi, which is the force of one pound acting on an area of one square inch:

$$1 \text{ psi} = 6890.16 \text{ pascal} = 6.89016 \text{ kPa}$$

$$1000 \text{ psi} = 6890.16 \text{ kPa} = 6.89016 \text{ MPa}$$

Another common pressure unit is inches of water, which is used in manometers, draft gauges, and differential pressure transducers. This is the pressure exerted by a column of water of a given height. Actually, it is the pressure exerted by the weight of that column of water, and this fact shows the difficulty of using inches of water as a pressure unit. The weight of the column of water will depend on the density of the water, which is influenced by the temperature of the water.

There are three common temperatures at which the inch-of-water unit is defined. These temperatures are:

> 4°C (39.2°F) — This is the temperature at which the density of water is at its maximum. At this temperature, 1 inch of water = 249.089 Pa.

60°F (15.56°C) — This is the standard reference temperature used by the American Gas Association for flowmeter calculations. At this temperature 1 inch of water = 248.836 Pa.

20°C (68°F) — This is the nominal room temperature. At this temperature, 1 inch of water = 248.641 Pa.

The difference in actual pressure between a 4°C reference and a 60°F reference is approximately 0.1%. We should always define the reference temperature when using inches of water as a pressure unit.

4.2.2 Strain Gauge Pressure Transducers

The strain gauge pressure transducer (see Figure 4-7) is the most common type of pressure transducer. Its operating principle is based on the deformation of a diaphragm when strain gauges directly or indirectly detect applied pressure. Strain gauges are constructed of fine wire or metal foil. The stretching of the wire or foil by the deformation of the diaphragm causes a reduction in the cross-sectional area of the wire or foil and a corresponding increase in the resistance. In a bonded strain gauge transducer, one or more strain gauges are attached directly to the diaphragm. In an unbonded strain gauge transducer, the deformation of the diaphragm is transmitted to another physical mechanism that contains the strain gauges. An example of this physical mechanism is four insulated posts with fine wire wrapped around them. The movement caused by the deformation of the diaphragm is coupled to the posts in such a way that they spread apart as pressure increases. This spreading stretches the wire and causes the resistance to increase.

The resistance change is measured by incorporating the strain gauge or gauges into a Wheatstone bridge circuit. This makes it possible to detect very small resistance changes, and also allows transducer manufacturers to add temperature compensating circuits to nullify resistance changes caused by temperature. The operating temperature range of each transducer will be limited by the materials used to bond the strain gauges, the solders used to connect wiring, and the type of wire insulation used. The pressure range of strain gauge transducers can vary from a few inches of water to thousands of psi. Low-range transducers require thin diaphragms with a large surface area in order

Figure 4-7. Strain Gauge Pressure Transducers
(Courtesy, Dynisco Instruments)

to generate sufficient strain in the diaphragm. High-range transducers can have thicker diaphragms with smaller areas.

The strain gauge bridge can be connected to a remote amplifier, or an integrated circuit amplifier can be built into the transducer housing. The remote amplifier arrangement allows for more flexibility when choosing the output signal and can allow the transducer to operate in a higher temperature environment. Its disadvantage is that electrical noise may be picked up on the long cable length. This noise, along with the strain gauge bridge output, will be amplified by the remote amplifier. The integrated circuit amplifier arrangement minimizes the noise problem since the output signal is at a high level as it leaves the transducer. Its disadvantage is that the temperature of the transducer environment is limited by the amplifier's electronics. The size of the transducer is also larger since it must accommodate the electronics required to amplify the signal.

4.2.3 Piezoelectric Pressure Transducers

In a piezoelectric pressure transducer, the sensing element is a material such as quartz crystal that produces a charge whenever the force exerted on it is changed. The word *piezo* is a Greek term meaning "to squeeze." When the force stops changing, the charge drains away or decays at a rate that depends on the material the element is made of and the time constant of the attached electronics. This electronics time constant is analogous to that of a first-order high-pass filter and is proportional to the resistance and capacitance of the filter. This decay means that piezoelectric pressure transducers are only used for dynamic measurements. Figure 4-8 illustrates a miniature version of a piezoelectric pressure transducer while Figure 4-9 illustrates a standard size transducer.

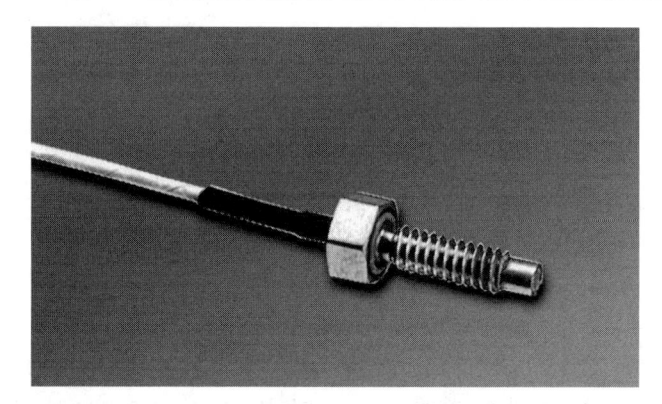

Figure 4-8. Miniature Piezoelectric Pressure Transducer
(Courtesy, Kulite Semiconductor Products, Inc.)

Figure 4-9. Piezoelectric Pressure Transducer
(Courtesy, Kulite Semiconductor Products, Inc.)

Piezoelectric pressure transducers are designed to be used with external amplifiers (charge mode transducers) or with integral electronic signal conditioning amplifiers. In either case, the amplifier's purpose is to convert the high-impedance output from the sensor to a low-impedance voltage output that can be coupled to displays, analyzers, and data acquisition systems. The advantage of the charge mode transducer is that it can operate in a high-temperature environment of up to

1000°F. The temperature environment of the integral amplifier transducer is limited to about 300°F because it uses microelectronic components. The high-impedance signal from the charge mode transducer can be affected by electromagnetic and radio frequency interference. It can also be affected by cable movement, which can create a charge between the conductors or between the conductor and shield. This charge, called the "triboelectric effect," is added to the charge produced by the transducer and creates an error signal.

To minimize the noise effects, special high-insulation resistance coaxial cable and connectors are used between the transducer and the amplifier. These cables and connectors must be kept clean and dry in order to maintain the high insulation resistance.

The triboelectric effect is minimized by using low-noise coaxial cable that has graphite lubricant between the shield and the dielectric surrounding the center wire. The integral amplifier transducer eliminates these problems since the output is a high-level, low-impedance signal. This signal can be transmitted with no degradation over standard coaxial or twisted-pair cable.

A compromise between these two systems can be obtained by using miniature in-line amplifiers. A short length of low-noise cable is run from the transducer to the in-line amplifier, which is located a safe distance from the high-temperature environment. Standard coaxial or twisted-pair cable is then run from the amplifier to the rest of the signal conditioning equipment. The choice of system you use depends on its cost, the transducer's environment, and the system's overall flexibility.

4.2.4 Piezo-Resistive Pressure Transducers

Piezo-resistive pressure transducers are based on the principle that semiconductor resistors change their resistance when they are deformed or stressed. The resistors are attached to a metal or ceramic diaphragm that deflects when pressure is applied. The resistors are usually incorporated into a Wheatstone bridge circuit similar to that used in the strain gauge transducer. Microelectronics technology is often used to manufacture these transducers so the resistors, temperature compensation circuits, signal conditioning amplifiers, and even the diaphragm can be integrated into a single unit. Unlike the piezoelectric transducer, the piezo-resistive transducer can make static as well as dynamic measurements.

4.2.5 Capacitive Pressure Transducers

The capacitive pressure transducer uses the properties of a parallel plate capacitor to measure the diaphragm motion that occurs when pressure is applied to the diaphragm. The capacitance varies when the distance between the plates is changed. In the transducer, one plate is connected to the diaphragm while the other plate is fixed.

4.2.6 Variable Reluctance Pressure Transducers

The variable reluctance pressure transducer is constructed with a magnetically permeable steel diaphragm and an inductance coil mounted on an E-shaped core. A small air gap is provided between the diaphragm and the legs of the core. This air gap provides a given reluctance for the magnetic flux path between the coil and the diaphragm. Pressure applied to the diaphragm reduces the air gap and changes the reluctance and hence the inductance of the coil. The coil is connected as one arm in an AC bridge, while a second coil acts as a reference for temperature compensation or as an active coil in a differential pressure transducer. In this case, the coils are located on opposite sides of the diaphragm, and, as the diaphragm moves, the inductance of one coil will increase while the other decreases. The other two arms of the AC bridge are usually made up of the center-tapped secondary of the transformer, which supplies the excitation voltage to the bridge. The AC output signal from the bridge is demodulated or rectified and filtered in order to produce a DC voltage signal that is proportional to pressure.

4.2.7 Resistive Pressure Transducers

This group of transducers can include the piezo-resistive transducers as well as an older style of pressure transducer in which the diaphragm deflection is mechanically amplified and then connected to the slider of a potentiometer. The potentiometer requires a simple signal conditioning system, consisting of a constant voltage or constant current supply connected across the potentiometer. The voltage that is measured between the slider and one end of the potentiometer is then proportional to the pressure.

4.2.8 Integrated Circuit Pressure Transducers

The integrated circuit pressure transducer uses microelectronics technology to produce the elements of a transducer on a chip. In some cases, even the diaphragm is constructed of the semiconductor material. Piezo-resistive and capacitive measuring techniques are often duplicated in the integrated circuit transducer. The signal conditioning amplifier and excitation voltage supply are contained in the same physical package as the sensing element. The advantages of the integrated circuit pressure transducer include its small size and low cost, while its disadvantages include its limitations on environmental temperature and upper pressure range. Typical low cost integrated circuit pressure transducers may only operate with environmental temperatures between 0 and 150°F and may only be able to measure up to 100 psig.

4.3 Flow Transducers

The flow transducers described in this section include differential pressure elements, turbine, vortex, thermal, Doppler, transit time, and laser Doppler anemometer.

4.3.1 Flow and Flow Measurement

Flow can be defined as the amount of material that is moved in a given time. The amount of material can be measured in mass units such as pounds and kilograms or in volume units such as gallons and cubic feet. The flow is known as mass flow if mass units are used and volumetric flow if volume units are used. If we know the density (mass per unit volume) of the material then we can convert between mass flow and volumetric flow. Using consistent units we get:

$$\text{Mass flow} = \text{volumetric flow} \times \text{density}$$

In most cases, the material is moving in a pipe or duct whose cross-sectional area is known. In a few cases, the material is moving in an open channel or on a conveyor belt. If the pipe is completely full, we can measure the flow by determining the velocity of the moving material. Using consistent units, we get:

$$\text{Volumetric flow} = \text{velocity} \times \text{cross-sectional area}$$

This relationship also requires either that the velocity be constant over the whole area or that it be possible to determine the velocity distribution (flow profile) by testing. Testing is done by dividing the area into small equal-area segments and then measuring the velocity in each segment. The average velocity is then calculated and used to determine the total flow. For similar conditions, a single velocity measurement at a known location can be adjusted to indicate the total flow. This procedure is known as a "traverse" since the velocity probe is often traversed across the pipe or duct. The pipe area is divided into a number of equal circumferential areas, while a duct is divided into a grid of equal areas. A value called the Reynolds number can be calculated to determine whether the flow profile will be flat (relatively equal over the pipe area) or parabolic (with higher velocity in the center of the pipe). The Reynolds number is based on the pipe area, the fluid viscosity, and the fluid velocity. For Reynolds numbers greater than 2300, the flow is called "turbulent" and has a flat profile. For numbers less than 2300, the flow is called "laminar" and has a parabolic profile. Many flow transducers are based on a velocity measurement, so it is important that we understand the flow profile effect.

4.3.2 Differential Pressure Flow Elements

Differential pressure flow measurements are based on the fact that a constriction in the pipe area will cause an increase in the fluid velocity and a corresponding reduction in the fluid pressure as the flow passes through the constriction. This pressure drop or "head" is measured using differential pressure transducers whose high-pressure port is connected to the pipe upstream of the constriction and whose low-pressure port is connected to the pipe downstream of the constriction.

The basic relationship between flow and pressure drop is that the flow is proportional to the square root of the pressure drop. Many other factors such as the type and size of constriction, the fluid's properties, and the locations of pressure taps are required to construct an equation that defines the exact relationship between flow and the square root of the pressure drop. Once all the factors are taken into account the relationship simplifies to this: flow is equal to a constant times the square root of the pressure drop. This equation will be valid as long as all the factors remain constant. In practice, the factors seldom remain constant. Thus, we take additional measurements such as fluid pressure and temperature and use these to adjust the flow equation in real time using some form of flow computer.

The most common type of restriction is the orifice plate. This is a flat plate one-sixteenth to one-quarter of an inch thick that has a hole bored through it that is smaller in diameter than the pipe's internal diameter. This plate is held between two pipe flanges so that the hole presents a narrowing of the flow path through which the fluid must pass. The most common form of the orifice plate is the concentric, sharp-edged orifice as shown in Figure 4-10. This orifice plate has the hole bored in such a way that the center of the hole is aligned with the centerline of the pipe. The upstream side of the hole is machined with a sharp edge, while the downstream side of thicker plates may be chamfered at an angle. Other designs include the eccentric orifice plate, which locates the hole above or below the pipe centerline; the segmental orifice plate, which has a d-shaped hole; and the quadrant-edged orifice plate, which has a quarter-round upstream edge. Each type of plate has been tested over a wide range of conditions, and a specific factor is included in the flow equation to account for the orifice type.

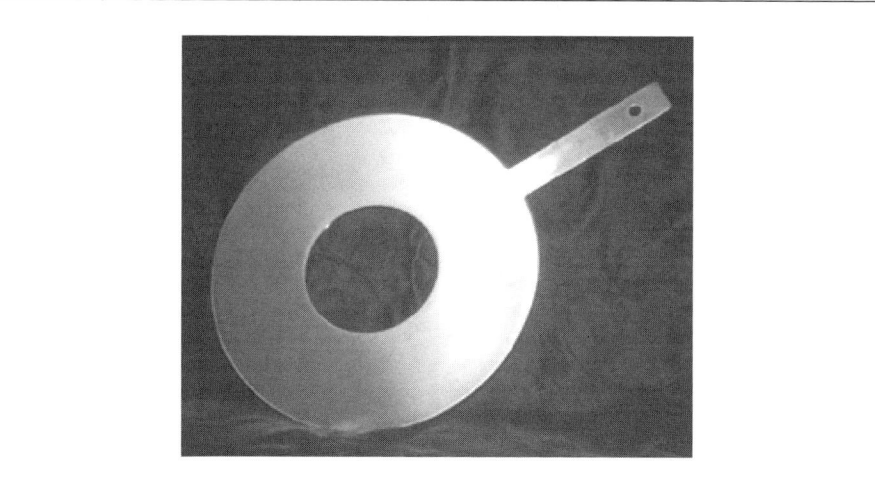

Figure 4-10. Concentric Bore Orifice Plate
(Courtesy, Fluidic Techniques)

The advantages of the orifice plate include its low cost and ease of installation. Its disadvantages include low rangeability (maximum to minimum flow) and high permanent pressure loss in the system, which is caused by the sudden restriction in flow area and the equally sudden return to full pipe area downstream of the plate's location. This permanent pressure loss requires a potentially expensive increase in the pumping capacity of the system.

Other types of restrictions include the venturi (see Figure 4-11), the flow tube (see Figure 4-12), and the flow nozzle (see Figure 4-13). These are restrictions whose constrictions have tapered inlets and that have a tapered or elongated outlet section. This design reduces the high permanent pressure loss associated with the orifice plate. The disadvantage of the venturi, flow tube, and flow nozzle is their higher cost.

Figure 4-11. Flow Venturi
(Courtesy, Fluidic Techniques)

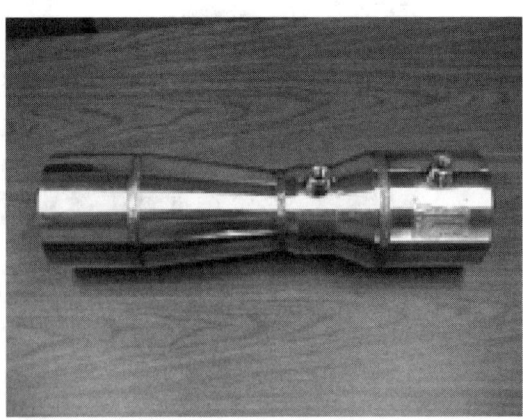

Figure 4-12. Flow Tube
(Courtesy, Fluidic Techniques)

Figure 4-13. Flow Nozzle
(Courtesy, Fluidic Techniques)

Several standards are available that detail the specific calculations and theory for each type of differential pressure flow element.[1]

4.3.3 Turbine Flowmeter

The turbine flowmeter has a series of angled blades mounted around a shaft that is positioned in the pipe so that the shaft is in line with the direction of flow. The flow impacts the vanes and spins the turbine at a rate proportional to the flow. The transducer is usually constructed with a body that supports the turbine shaft on bearings. The turbine blades are constructed so that their tips are very close to the inside walls of the flow passage through the body. Figure 4-14 shows the internal construction of a typical turbine flowmeter. This body is provided with pipe flanges to mate with flanged pipe sections or is designed to be clamped between a set of pipe flanges. An inductive pickup is located in the body directly above the blade tips. This probe outputs a pulse whenever a blade cuts through its magnetic field. Since

---

1. See International Organization of Standards (ISO 5167-1), 1991: "Measurement of fluid flow by means of pressure differential devices, Part 1: Orifice plates, nozzles, and Venturi tubes inserted in circular cross-section conduits running full." Reference number: ISO 5167-1:1991(E). See also, International Organization of Standards (ISO 5167-1) Amendment 1, 1998: "Measurement of fluid flow by means of pressure differential devices, Part 1: Orifice plates, nozzles, and Venturi tubes inserted in circular cross-section conduits running full." Reference number: ISO 5167-1:1991/Amd.1:1998(E). Also see, American Society of Mechanical Engineers (ASME). 1971, *Fluid Meters: Their Theory and Application*. Edited by H. S. Bean. 6th ed. Report of ASME Research Committee on Fluid Meters.

it takes a specific volume of fluid to rotate the turbine past the pickup, the manufacturer can provide a calibration constant that gives the number of pulses per volume of fluid. The signal conditioning electronics determine the number of pulses over a given time and outputs a signal that is proportional to the volumetric flow rate (gallons per minute or cubic feet per minute). The turbine flowmeter can provide a ten-to-one maximum to minimum flow range, as compared to a four-to-one range for the orifice plate. Figure 4-15 illustrates turbine flowmeters with a variety of sizes and pipe attachments.

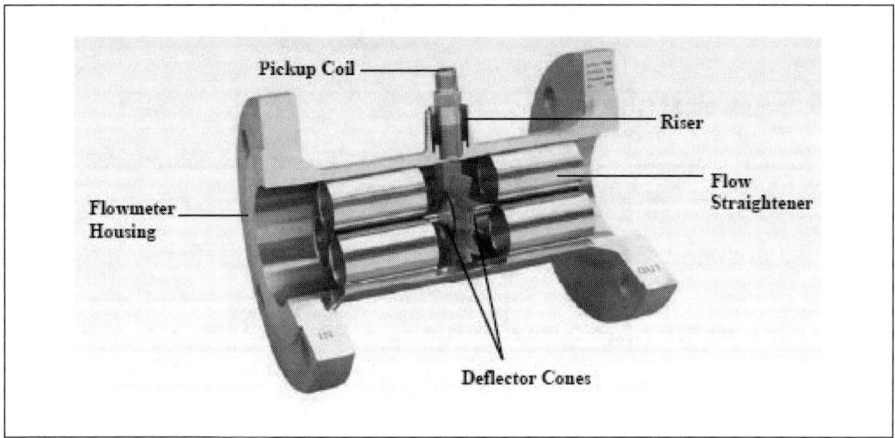

Figure 4-14. Turbine Flowmeter Internal Details
(Courtesy, Hoffer Flow Controls, Inc.)

Figure 4-15. Turbine Flowmeters
(Courtesy, Hoffer Flow Controls, Inc.)

4.3.4 Vortex Flowmeter

The vortex flowmeter (see Figure 4-16) is made by inserting a bluff body (i.e., having a nonstreamlined shape) into the fluid stream. A bluff body has an upstream face of a certain area that obstructs the flow. The body size tapers down to produce a smaller downstream area. This shape causes the flow to produce vortices or eddies that appear on alternate sides of the bluff body at a frequency that is proportional to the flow. This phenomenon is known as "vortex shedding." As the vortices separate from the trailing edge of the bluff body, they create a pressure change in the fluid that can be detected with piezoelectric transducers. Other transducers that have been used to detect the oscillation frequency include strain gauges and thermistors.

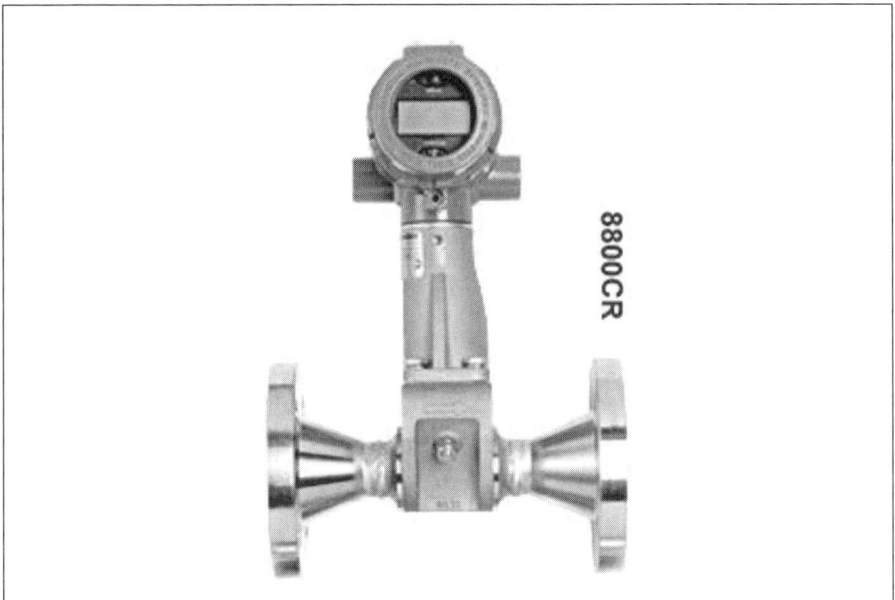

Figure 4-16. Vortex Flowmeter
(Courtesy, Emerson – Rosemount)

The vortex flowmeter typically has a rangeability of ten to one or even twenty to one. Accuracy can range from ½ to 1 percent of flow rate. The vortex flowmeter has a smaller permanent pressure loss than the orifice plate and does not suffer from the flow-induced wear effects that can degrade the accuracy of the orifice plate.

4.3.5 Thermal Flowmeter

The thermal flowmeter typically consists of two probes mounted into the flow path. The upstream probe contains a temperature sensor to monitor the temperature of the probe. This temperature will be directly related to the temperature of the fluid. The downstream probe contains a heater as well as a temperature sensor. Under no-flow conditions the temperatures of the two probes will differ. When flow is increased, the fluid will draw heat away from the downstream probe, and the temperature difference will decrease. The amount of heat removed from the downstream probe is proportional to the mass flow rate of the fluid.

4.3.6 Doppler Flowmeter

The Doppler flowmeter (see Figure 4-17) uses the frequency shift in an ultrasonic sound wave that is transmitted into a flowing fluid, reflected from particles in the fluid moving at the fluid's velocity, and received at a receiver for processing. The frequency shift is proportional to the fluid's velocity. This change in frequency or Doppler shift

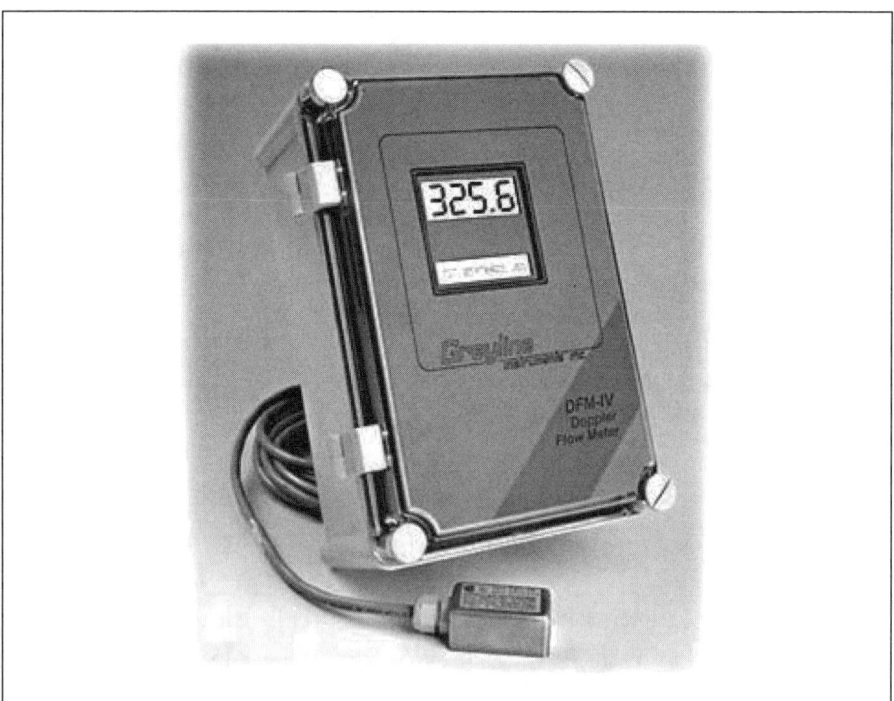

Figure 4-17. Doppler Flowmeter
(Courtesy, Greyline Instruments Inc.)

occurs whenever the source of the sound wave is moving in relation to the receiver or the receiver is moving in relation to the source. We are all familiar with the sound of a train's horn being higher pitched (higher frequency) as the train approaches us and then instantly dropping to a lower pitch as the train passes us and travels in the opposite direction.

Some Doppler flowmeters have a transmitter and receiver mounted on opposite sides of the pipe, while in others the transmitter and receiver are combined into a single transceiver. The transceiver must be acoustically coupled to the fluid, which means that there must be a continuous path through which the sound wave can travel. Permanently installed flowmeters can have the transceiver protruding through the pipe wall into the fluid. This is called a "wetted transducer." Portable flowmeters have the transceiver clamped to the outside of the pipe wall with a coupling fluid (gel or paste) filling in any gaps between the transceiver face and the wall of the pipe.

The Doppler flowmeter requires that particles or entrained air bubbles be present in the fluid to reflect the sound wave from transmitter to receiver. These reflectors must be of some minimum size and be present in some minimum concentration for the meter to operate. The frequency shift also depends on the speed of sound in the fluid. The speed of sound will vary according to the fluid's density and pressure, and most meters will have programmable inputs for selecting the fluid type or the expected speed of sound.

The Doppler flowmeter measures fluid velocity. It requires a flat flow profile in order to convert velocity into flow units using the inside area of the pipe. This means that the transceiver must be installed on a straight run of piping located some distance from pipe elbows and valves that can distort the flow profile.

The advantages of Doppler flowmeters include the capability of making measurements from the outside of the pipe and the lack of permanent pressure loss that they make possible, even with wetted transducers. Their disadvantages include accuracies limited to 1 to 2 percent, the fact that they can only be used with transducers limited to about 400°F, and their inability to operate on very clean fluids like deionized water.

4.3.7 Transit Time Flowmeter

The transit time flowmeter (see Figure 4-18) also uses ultrasonic sound waves to determine the fluid's velocity. Upstream and downstream transceivers alternately send and receive a burst of ultrasonic energy. The time it takes for the pulse from the upstream transceiver to reach the downstream transceiver will be less than for the opposite direction. The difference in these times is proportional to the fluid's velocity.

Figure 4-18. Transit Time Flowmeter
(Courtesy, Controlotron Corp.)

The transit time ultrasonic flowmeter works best on clean fluids since particles and bubbles will attenuate the intensity of its ultrasonic beam. If straight pipe length is at a minimum, the transceivers should be placed on opposite sides of the pipe and spaced a distance of approximately one half of the pipe diameter apart. This configuration is also used whenever conditions that would attenuate the signal are present, such as fluids with high turbidity or pipes that have linings or corroded inner walls.

The optimum mounting method is to mount the transceivers on the same side of the pipe and to bounce the signal from one transceiver off the opposite wall of the pipe and onto the other transceiver. A third method, used for small-diameter pipes, is to bounce the signal off both walls of the pipe so that the path follows the shape of a W. This provides a longer path and increases the transit time to an amount that that can be more accurately measured.

Clamp-on transceivers are usually mounted on a rail that is clamped along the length of the pipe. The transceivers can be slid along these rails and fixed in the position that provides the optimum signal reception. The flowmeter's microprocessor-based signal conditioning can determine this spacing by using pipe data entered by the user and then refining the spacing by performing trials on a no-flow condition.

As with the Doppler flowmeter, the user must be aware of both upstream and downstream piping components and the piping configurations that could affect the flow profile.

4.3.8 Laser Doppler Anemometer

The Laser Doppler Anemometer (LDA) or Laser Doppler Velocimeter (LDV) is a laboratory flow measurement system that measures fluid velocity noninvasively. No components are inserted into the flow path that could disturb the flow distribution.

The system consists of a single laser beam that is split into two beams of equal intensity. These two beams are directed at a common point in the flow path from two different angles. An interference pattern of light and dark bands is created within the football-shaped volume that is formed at the point at which the two beams intersect. Particles in the fluid move through this measurement volume and reflect light when they pass a light band. This reflected light is collected by a photo detector. The frequency of the photo detector's output as the particle passes from one light band to another is proportional to the particle's velocity. The system assumes that the particle is moving through the volume at right angles to the light and dark bands. If not, it will only report the component of the velocity that is moving in this direction.

If another laser beam pair is directed at the same measuring volume from a different angle, then two components of the velocity can be established. A third laser beam pair can provide a third component of the fluid velocity. A three-component system can detail the velocity and direction of flow in a very small part of the total flow volume. By moving the lasers to various points in the system, we are able to develop a picture of the flow distribution throughout the flow path. Determining this distribution around components in a flow is often the goal of LDV users.

These systems are expensive and are generally limited to use in research and development laboratories. They require a clear window into the fluid and the presence of light reflecting particles. The test fluid can often be seeded with such particles if they are not normally present.

Table 4-3 summarizes the advantages and disadvantages of the various types of flowmeters.

Table 4-3. Comparison of Flow Sensors

Sensor Type	Advantages	Disadvantages
Differential Pressure	Fluid or gas measurement Mature technology	High permanent pressure loss Upstream and downstream piping effects Many fluid-dependent effects Low rangeability
Turbine	High accuracy High rangeability Small piping effects	High permanent pressure loss Liquid must be clean Expensive
Vortex	Medium rangeability Good accuracy	Medium pressure loss Expensive
Thermal	Fluid or gas measurement Mass flow measured	
Doppler	No permanent pressure loss Can be portable	Temperature limitation No ultraclean fluids
Transit Time	No permanent pressure loss Can be portable	Temperature limitation Works best on clean fluids
LDA	Nonintrusive Measures velocity in small area	Expensive

4.4 Displacement Transducers

Displacement transducers are used to measure the physical distance between the sensor and a target. Since most displacement transducers can measure dynamic as well as static displacements they are often used to measure the vibration of a target. The range of displacements measured can range from a few micro inches, using laser-based transducers, to tens of feet, using cable-driven potentiometers. In some types, there must be physical contact between the target and some part of the transducer. Other transducers do not require any contact.

4.4.1 Linear Displacement and Linear Displacement Measurement

Linear displacement transducers measure the dimensional change that occurs on a straight line between the transducer and the target. Their relevant units of measurement include the inch and foot as well as the metric units of centimeter, millimeter, and micrometer. The measurement system has the usual components: a transducer, cabling, and a signal conditioner. The typical system produces a voltage output that is linearly proportional to the distance measured.

4.4.2 Linear Variable Differential Transformer (LVDT)

The Linear Variable Differential Transformer or LVDT consists of a movable magnetic rod that acts as a core element between the primary coil and two secondary coils that surround the rod. The LVDT's body is usually cylindrical, with the core moving in a central, tubular cavity that runs the length of the cylinder. The primary coil is wound around the tube at the center of the cylinder, while the secondary coils are wound around the tube at opposite ends of the cylinder. Figure 4-19 illustrates the construction details of a typical LVDT. The primary coil is connected to a constant voltage AC source that typically supplies 3 volts rms at a frequency of 3 kilohertz. The magnetic flux that the primary coil develops is coupled into the secondary coils by the movable rod. How much flux is coupled into each coil depends on the position of the rod in relation to the coils. As the core moves toward one coil the amount of flux coupled into that core increases, as does the output voltage. At the same time, the output of the other coil decreases. Over a certain range of movement, the difference in these voltages is proportional to the movement of the rod. The secondary coils are connected in series opposition so that when the rod is centered between the two coils, the flux developed in each is equal but the output voltages are canceled out due to the 180-degree phase shift. This position is known as the "null position." The signal conditioning uses the phase shift that occurs at the null position to determine the direction of the rod's movement.

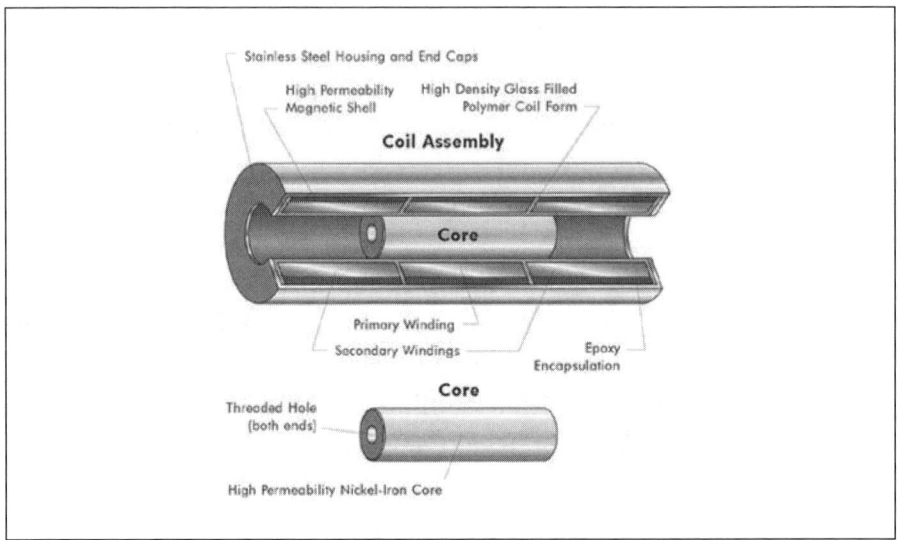

Figure 4-19. LVDT

The electronics in the signal conditioner is known as a "phase-sensitive demodulator"; it changes the differences in AC voltage to an equivalent DC voltage. The phase-sensing circuitry determines when the rod passes through the null point and changes the polarity of the DC output signal. A typical output signal will be +10 volts at one extreme of the rod movement, 0 volts at the null position, and -10 volts at the other extreme of the rod position. Figure 4-20 illustrates the operation of an LVDT.

An LVDT that uses external signal conditioning is known as an AC LVDT, while one in which microelectronic circuitry is built into the cylindrical housing is known as a DC LVDT. The DC LVDT only requires the addition of a DC power supply to create a complete measurement system.

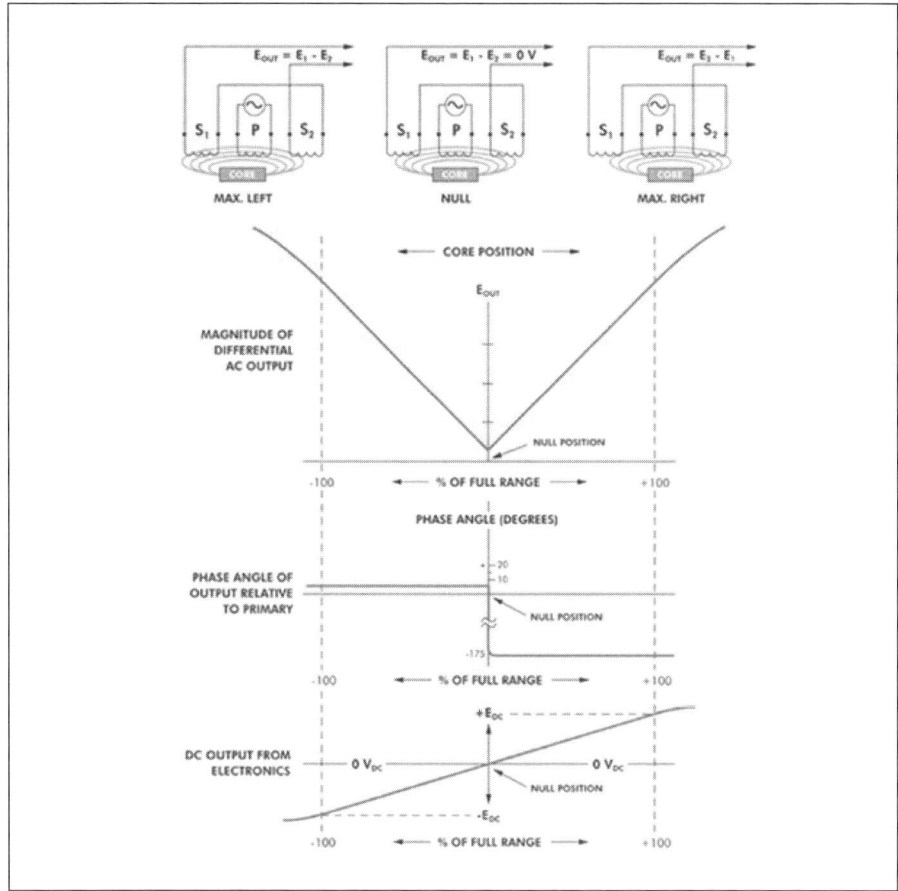

Figure 4-20. LVDT Operation

4.4.3 Capacitive Displacement Transducer

Capacitive displacement sensors measure the capacitance between the probe face and the surface of the conductive target. The probe and target form a basic parallel plate capacitor. The capacitance varies directly with the plate area and the dielectric constant of the material between the plates. It varies indirectly with the gap between the plates. Since the plate area is fixed by the probe diameter and the dielectric is a constant (air), then the capacitance is only dependent on the gap. Figure 4-21 shows examples of capacitive displacement probes.

Normally, the changing electric field that is applied to the probe face would spread in all directions, and errors would occur if the field encountered conductive material other than the target. This is prevented by surrounding the sides and back of the probe with a guard, which is a conductor driven at the same voltage as the probe but using

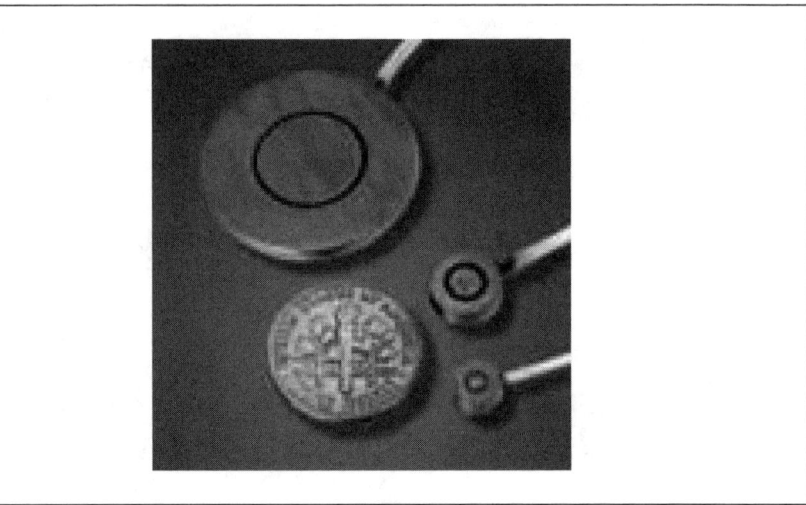

Figure 4-21. Capacitive Displacement Transducers
(Courtesy, Capacitec Inc.)

a separate source. Since the probe and guard are at the same voltage, no field exists between them. Any conductors around the probe will form a field with the guard and not with the probe. This tends to focus the field onto the target rather than the unguarded front of the probe. A round probe will focus a cylindrical field onto the target. The diameter of this field will spread to be approximately 30% larger than the probe diameter. The target diameter has to be at least this size. The maximum gap that can be measured is about 40% of the probe diameter, although most sensors are limited to considerably less than that. This means that for larger gap measurements, the diameter of the probes must increase. Figure 4-22 illustrates the construction details of a typical capacitive displacement transducer.

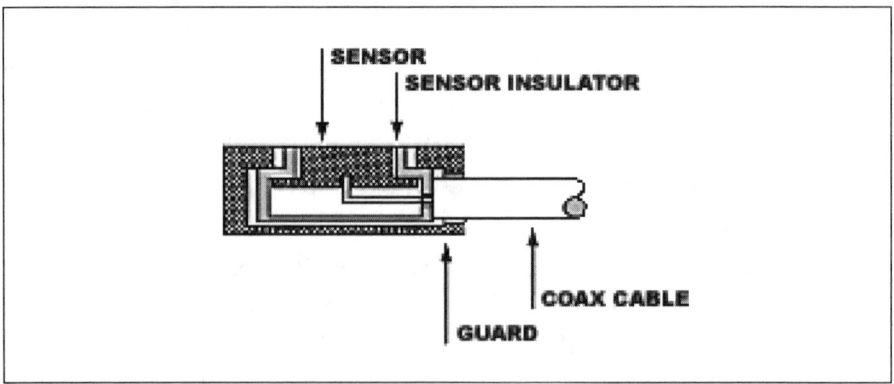

Figure 4-22. Capacitive Displacement Transducer Construction
(Courtesy, Capacitec Inc.)

The advantages that capacitance displacement transducers offer are noncontact measurement and high sensitivity. Typical systems will provide a sensitivity of 1 volt per 100 micrometers. Their disadvantage is their limited range of measurement, even with a reasonably sized probe.

4.4.4 Inductive Displacement Transducer

Inductive displacement transducers detect the presence of a metallic target by using a coil that is driven by a high-frequency oscillator. The coil sets up a magnetic field that radiates from the face of the transducer. When a target enters this magnetic field, eddy currents are created in the target material. The energy required to generate the eddy currents is removed from the coil oscillator, and the amplitude of the oscillator voltage decreases. The closer the target is to the transducer face, the more the oscillator amplitude is decreased.

4.4.5 Laser Displacement Transducer

Laser displacement transducers use the principle of triangulation to measure linear displacement. The transducer focuses the light from a laser diode onto a small spot on the target's surface. The transducer also has a receiving lens that is set at an angle to the transmitted beam. The reflected light that bounces off the target and enters this lens is focused onto a certain position on a light-sensitive device, such as a photodiode array or a charge-coupled device. If the target moves to another location, the reflected light will be focused onto a different position on the light-sensitive device. Figure 4-23 shows the operation of a typical laser displacement probe.

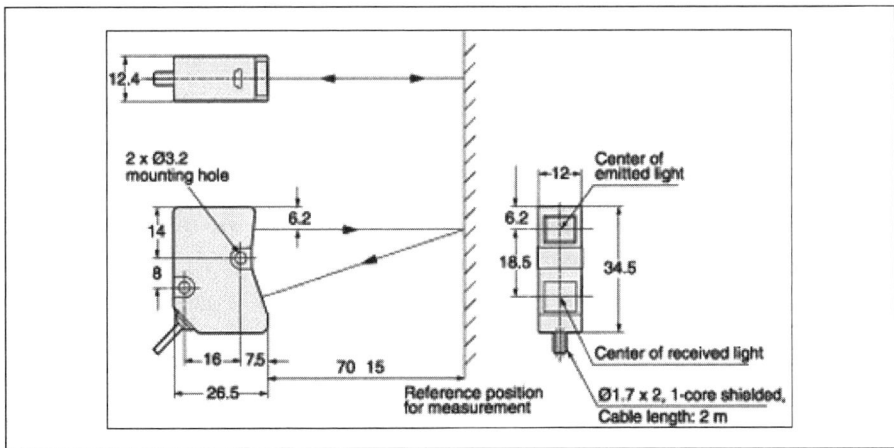

Figure 4-23. Laser Displacement Transducers
(Courtesy of Keyence Corporation)

4.4.6 Resistive Displacement Transducer

The resistive displacement transducer in its most basic form is a linear potentiometer whose slider is attached to or spring loaded against the target. As the target moves, the wiper will be located to a new position on the resistance element. These devices are inexpensive and require simple signal conditioning, consisting of a DC power supply and a voltmeter. The disadvantages of the resistive displacement transducer are that it needs to be in contact with the target and if the wiper travel is excessive either in speed or duration the resistive element becomes worn.

Another type of resistive displacement transducer is the cable extension transducer. This transducer measures the drum's rotation using a thin cable wound on a spring-loaded drum and a multi-turn potentiometer. The end of the cable is attached to the target. If the target moves in a linear manner the drum and potentiometer will rotate in a proportional motion.

4.4.7 Ultrasonic Displacement Transducer

Ultrasonic displacement transducers emit an ultrasonic sound wave that is reflected from the target. The distance between the target and the transducer can be calculated by measuring the time it takes the wave to reflect and the wave's propagation velocity in the medium between the probe and target. This principle can be applied to ultrasonic thickness transducers, which measure the thickness of materials or coatings. In these types, the ultrasonic wave is coupled into the material by coating the probe's face with a gel and pressing it against one surface of the material. The signal conditioning is able to detect the time required for part of the wave to reflect from the back face of the material to the front face. Figure 4-24 illustrates a typical ultrasonic displacement transducer.

Table 4-4 summarizes the advantages and disadvantages of the various types of linear displacement transducers.

Figure 4-24. Ultrasonic Displacement Transducer
(Courtesy, Senix Corporation)

Table 4-4. Comparison of Linear Displacement Sensors

Sensor Type	Advantages	Disadvantages
LVDT	Many measurement ranges Infinite resolution	Long sensor length Core attached to target
Capacitive	Noncontact High output sensitivity	Small measurement range Temperature sensitive
Inductive	Many measurement ranges Some types are noncontact	
Laser	Noncontact Fast response Wide frequency response	Expensive Clean environment required
Resistive	Inexpensive Many measurement ranges	Requires contact with target Low frequency response Mechanical wear
Ultrasonic	Noncontact Fast response High resolution	Temperature sensitivity Expensive

4.4.8 Rotational Displacement and Rotational Displacement Measurement

Rotational displacement is the distance a target moves in a circular path. Transducers are made to measure some part of a complete circle, a complete circle, or multiple rotations around the same circular path. The units of rotational displacement include degrees, minutes, and seconds. A complete circle contains 360°, while 1° contains 60 minutes and 1 minute contains 60 seconds. Rotational displacement transducers can also be used to measure the angle through which a target has moved.

4.4.9 Rotational Variable Differential Transformer (RVDT)

The Rotational Variable Differential Transformer (RVDT) is the rotational version of the LVDT. Its coils are contained within a cylindrical body that is shaped much like a multi-turn potentiometer or a small motor. The RVDT's core is connected to a shaft that protrudes from the end of the cylinder. The shaft rotates the core, which is shaped to couple more or less of the electromagnetic flux from the primary coil to the secondary coils. The RVDT's output signal is linearly proportional to shaft rotation over a range of approximately ±40° about the null point.

The RVDT's disadvantages are its limited rotational range and the fact that its shaft must be physically coupled to the target. Its advantage is the infinite resolution of the output signal, which means it can detect very small rotational movements.

4.4.10 Potentiometers

Rotary potentiometers are commonly used in electronic circuits to provide a variable resistance that is proportional to shaft rotation. The variable resistance is used to cause some change in the electronic circuit such as volume control in a radio or speed control of a motor. When used as a transducer, the potentiometer's shaft is coupled to the target, and the resistance change is used to determine the target rotation. The resistive element in the potentiometer can be made of a helix or spiral of fine wire or of a resistive film deposited on a substrate. In either case, shaft rotation causes a mechanical wiper contacting the element to move in a manner that increases or decreases the resistance between the wiper and one end of the element.

In the case of the wire-wound potentiometer, the wiper moves from one wire to the adjacent wire with a corresponding step change in resistance. This step change, while small, determines the minimum rotation that the potentiometer is capable of measuring. The resistive film potentiometer essentially has infinite resolution since the resistance change is based on the size of the particles making up the resistive paste that is used to construct the resistance element.

The potentiometer can be manufactured to measure part of a full rotation, almost a full rotation, or multiple rotations.

Table 4-5 summarizes the advantages and disadvantages of the various types of rotational displacement transducers.

Table 4-5. Comparison of Rotational Displacement Sensors

Sensor Type	Advantages	Disadvantages
RVDT	Infinite resolution	Limited rotational range Requires core attachment
Potentiometers	Inexpensive Single turn or multi-turn	Low frequency response Mechanical wear
Inclinometer	High resolution	Some are fragile

4.4.11 Inclinometers

An inclinometer or tilt sensor is a transducer that is designed to measure the number of degrees that a target is oriented away from being vertical. This means that these transducers use the earth's gravitational force to establish a vertical reference that is a line between the transducer and the center of the earth.

One type of inclinometer uses a pendulum mounted on the shaft of a rotary displacement transducer such as an RVDT. As the transducer mounted on a target is tilted, the mass on the end of the pendulum arm is forced by gravity to align itself vertically. This rotation is measured by the RVDT, which determines the number of degrees of tilt in one plane or one axis. A second inclinometer mounted at a 90° angle to the first inclinometer is required to determine the tilt in the other axis.

A second type of inclinometer is based on the bubble level that carpenters use to determine a vertical alignment. This transducer uses a curved glass vial filled with an electrically conductive fluid, a bubble of air, and electrodes that penetrate the glass and contact the liquid. This type of inclinometer is shown in Figure 4-25. When the transducer is level, the bubble is at the top center of the curved vial, and a certain amount of liquid is located between electrodes. The resistance of this amount of liquid is measured, and this resistance represents zero degrees of tilt. As the vial is tilted the air bubble moves to remain at the top of the vial and displaces some of the liquid between certain electrodes, thus changing the resistance. Transducers can be made with two vials in one housing so that tilt in two axes can be measured.

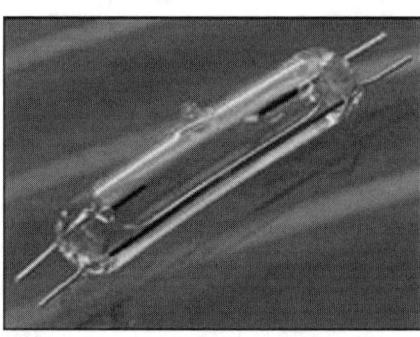

Figure 4-25. Inclinometer
(Courtesy, The Fredericks Company)

Low range versions of the integrated circuit accelerometers described in section 4.6.5 can also be used as inclinometers. Units with ranges of ± 1 g to ±2 g will respond to the earth's gravity, which has a nominal value of 1 g (see Figure 4-26). Integrated circuit accelerometer units have a DC response. They will change their output if they are tilted since only the component of the gravity field that acts in a normal direction relative to the accelerometer mass and capacitor plates will change the output. The output of these sensors will be proportional to the sine of the tilt angle (see Figure 4-27). There are two axis versions of these sensors in which two accelerometers are mounted in planes that are different by 90°. These versions can be used to determine the tilt angle of a target in two axes.

Figure 4-26. Inclinometer
(Courtesy, Instruments & Control, Inc. www.singer-instruments.com)

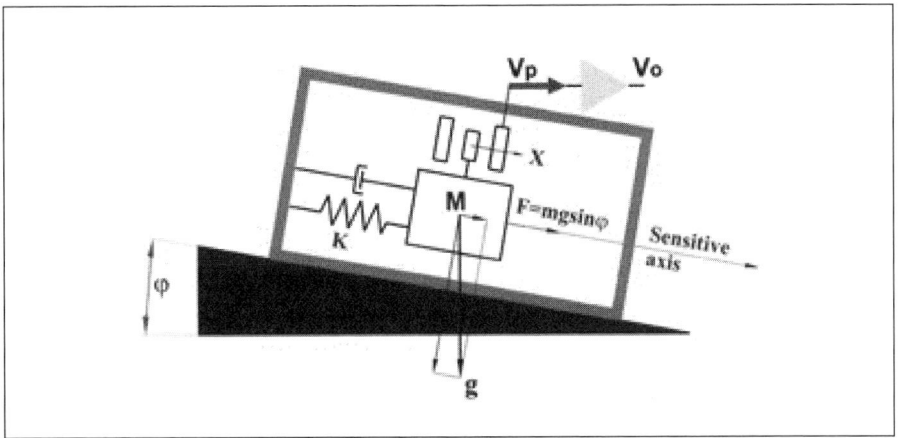

Figure 4-27. Inclinometer Operation
(Courtesy, Instruments & Control, Inc. www.singer-instruments.com)

4.5 Velocity Transducers

The types of velocity transducers discussed in this section use the piezoelectric, strain gauge, and electrodynamic methods of operation.

4.5.1 Velocity and Velocity Measurement

Velocity is a measure of how the displacement is changing with respect to time. Velocity is a vector quantity and thus has a magnitude and a direction. The speedometer in our car tells us how fast we are going but doesn't tell us in which direction. Speed is the magnitude of the velocity.

The units of velocity include inches per second and feet per second for linear velocity transducers and degrees per second or radians per second for rotary transducers.

Linear velocity transducers are used mainly to monitor vibration of rotating machinery such as motors, generators, and pumps.

4.5.2 Piezoelectric Velocity Transducers

Piezoelectric velocity transducers are piezoelectric accelerometers (see section 4.6.2) with built-in electronic integrators to provide an output that is proportional to velocity.

4.5.3 Strain Gauge Velocity Transducers

Strain gauge velocity transducers are strain gauge accelerometers (see section 4.6.3) with built-in electronic integrators to provide an output that is proportional to velocity.

4.5.4 Electrodynamic Velocity Transducers

Electrodynamic velocity transducers use the movement of a permanent magnet relative to a coil of wire to produce an output that is proportional to the velocity of the movement. In some designs, the magnet is mounted on a rod that moves through a hollow core around which the coil is wound. In other designs, the coil is suspended on springs and attached to a dampener within the transducer body while the magnet is fixed to the body. The springs and dampener allow the coil to remain in a fixed position in space while the transducer body and magnet move.

The rotary version of the velocity transducer is the tachometer in which a permanent magnet revolves within a coil.

As the name implies, electrodynamic velocity transducers only provide an output while the magnet is moving in relation to the coil. The typical frequency response range of these velocity transducers is from 10 hertz (cycles per second) to 1000 hertz. This is a good match for the frequency of the vibrations typically associated with rotating machinery.

4.6 Acceleration Transducers

Acceleration transducers or accelerometers use piezoelectric, strain gauge, variable reluctance, or integrated circuit technology as basic principles of operation.

4.6.1 Acceleration and Acceleration Measurement

Acceleration is a measure of how the velocity is changing with respect to time. Acceleration is a vector quantity and thus has a magnitude and a direction. The units of acceleration include inches per second per second (inches per second squared) or feet per second per second (feet per second squared). Another unit of acceleration is the g, which is the acceleration of a body due to the force of the earth's gravity. One g is equal to 32.2 feet per second squared or 9.81 meters per second squared.

Transducers that are used to measure acceleration are normally called accelerometers. They are used most commonly to measure the vibration of an object.

4.6.2 Piezoelectric Acceleration Transducer

Piezoelectric accelerometers (see Figure 4-28) are constructed using a mass that is spring loaded against a crystal. Accelerating the transducer will increase or decrease the force on the crystal, depending on the direction of the acceleration. The magnitude of the force is equal to the mass times the acceleration (F = ma). The force acting on the crystal will produce an electrical charge across opposite faces of the crystal that is proportional to the applied force. We thus have a transducer that produces a charge output that is proportional to acceleration.

Figure 4-28. Piezoelectric Accelerometer
(Courtesy, Brüel & Kjær)

As with the piezoelectric pressure transducer, the charge will leak away once acceleration ceases to change. The piezoelectric accelerometer is a dynamic transducer that is only used to measure acceleration that is changing with respect to time.

It is even more essential to use low-noise cable when using accelerometers since target vibration will definitely be present. The cable should be fixed to the target surface as close as possible to the transducer.

Piezoelectric accelerometers are manufactured in a wide variety of shapes and sizes to meet the requirements of specific tests. For example, extremely small transducers with a small mass are required when making measurements on light or fragile components. The mass of these transducers should not be greater than 10% of the dynamic mass of the test structure if the goal is to keep vibration levels and frequencies relatively unchanged when adding the transducer to the structure. Transducers whose internal masses are large enough to provide sufficient force on the crystal are used for very low-level acceleration measurements. Triaxial accelerometers (see Figure 4-29) have three accelerometers mounted in the same body. These three accelerometers are mounted in the x, y, and z planes to provide an output in each axis.

Figure 4-29. Triaxial Piezoelectric Accelerometer
(Courtesy, Brüel & Kjær)

The low frequency limit of operation is determined by the amount of charge leakage from the transducer, cabling, and signal conditioner and by the effects of ambient thermal variations on the transducer's sensitivity. In practice, the signal conditioning amplifier has a high-pass filter to cut off low frequency signals at a known frequency. The upper frequency limit is determined by the resonant frequency of the transducer's mass-spring system. The accelerometer's sensitivity will increase dramatically as this frequency is approached. In general, the usable upper frequency should be limited to approximately one-third of the resonant frequency. A low-pass filter is used in the signal conditioning amplifier to limit the output signal to frequencies that are well below the resonant frequency.

The method by which the accelerometer is mounted to the vibrating target is extremely important since we want the accelerometer to faithfully follow the target's motion at all frequencies. Poor mounting techniques will produce a mounted resonant frequency that is lower than the accelerometer's resonant frequency. The best mounting method is to drill and tap into a smooth, flat surface and mount the accelerometer tightly to the surface using a threaded steel stud. The manufacturer will provide wrench flats on the accelerometer and a suggested mounting torque. If drilling and tapping into the target is impractical then studs with a wide base can be cemented to the target using a glue that sets or dries to a hard material. A thin layer of beeswax to bond the accelerometer to the target can provide a less permanent mounting method. Beeswax should only be used in applications involving a temperature of approximately 100°F or less. Magnetic mounting bases are available, but these tend to lower the mounted resonant frequency much more markedly than do the previously described methods.

4.6.3 Strain Gauge Acceleration Transducer

The strain gauge accelerometer uses a mass-and-spring assembly to transfer the force created by the mass's acceleration to a diaphragm or beam that is instrumented with strain gauges. The two forms of these strain gauges are the metal foil type and the semiconductor type. The strain gauge circuits and signal conditioning are similar to those used in strain gauge pressure transducers.

The strain gauge accelerometer offers the advantage of having a frequency response that extends to DC or steady state. The size and mass of the transducer are typically larger than those of the piezoelectric accelerometer.

4.6.4 Variable Reluctance Acceleration Transducer

The variable reluctance accelerometer is typical of a class of accelerometer that determines acceleration by measuring the movement of a spring-suspended mass. The mass is connected by the springs to the accelerometer case, and a motion dampener is added to prevent overshoot and oscillations caused by quick changes in motion. When the case is accelerated, the mass tends to remain in position due to inertia. When the acceleration stops, the springs return the mass to the zero position. The relative displacement of the mass and case is proportional to the acceleration of the case.

In the variable reluctance accelerometer, the mass carries a permanent magnet that moves in proximity to a coil of wire. The magnet movement induces a voltage in the coil as long as the movement continues.

This type of accelerometer is limited to low-frequency measurements because of the low resonant frequency of the mass-spring assembly. They are typically used to measure seismic activity or in oil exploration testing.

4.6.5 Integrated Circuit Acceleration Transducer

Integrated circuit accelerometers employ micromachining technology to place a mass-spring type accelerometer and the signal conditioning electronics on an integrated circuit chip. The mass movement is measured by parallel plate capacitors.

These transducers are available in one-, two-, or three-axis versions with ranges from ± 1g to ± 100g.

Table 4-6 summarizes the advantages and disadvantages of the various types of acceleration transducers.

Table 4-6. Comparison of Acceleration Sensors

Sensor Type	Advantages	Disadvantages
Piezoelectric	Small size and mass High-frequency response High sensitivity Wide operating temperature range	Only AC response
Strain gauge	DC and AC response	Larger body size
Variable reluctance		Low-frequency response
Integrated circuit	Small size and mass Inexpensive Simple signal conditioning	Limited upper temperature limit

4.7 Force Transducers

Force transducers are constructed in a wide variety of shapes and sizes. The basic types include load cells, load buttons, and load washers.

4.7.1 Force and Force Measurement

Force can be defined as the condition or influence that attempts to change the speed and direction of a body that is stationary or already moving. This change in speed or in velocity is the acceleration that the body undergoes as a result of the force being applied to it. The direction of the acceleration is the same as the direction of the applied force.

The basic equation describing force is as follows:

$$\text{Force} = \text{mass} \times \text{acceleration}$$

In more general terms, force is the amount of pushing or pulling that is being applied to an object. Force sensors or load transducers are used to measure this pushing or pulling. To measure the applied force, the load transducer must be located between the object and the source of the force. When pushing, the transducer is said to be compressed or to be in compression. When pulling, the transducer is being stretched or is in tension. Some load transducers only measure compression loads, while others measure only tension loads. Universal load transducers measure both compression and tension loads.

Load transducers operate by measuring the strain that is produced in a material as a result of an applied force. This force applied over a unit area is known as the stress. At low stress levels, the material will change shape as a result of the stress but will return to its original shape when the stress is removed. Strain is a measure of the percentage change in the length of the material as a result of the applied stress. The majority of load transducers use some form of strain gauge as the basic measurement element.

A common use of load transducers is to measure the weight of an object. Weight is the force applied to a mass as a result of the acceleration caused by gravity.

The units of force or load include the pound and the kilogram. These should properly be called the pound-force and the kilogram-force to distinguish them from the units of mass. The more scientifically correct units include the slug and the Newton.

The most important concerns when using load transducers are to ensure that the applied load does not exceed the limit specified by the manufacturer and that the load is applied only to the sensitive axis of the transducer.

4.7.2 Load Cells

Load cell is the generic name given to a wide variety of load or force transducers. Other names have been given to force transducers that have a specific shape or carry out a specific type of force measurement. Figure 4-30 illustrates a typical load cell.

Figure 4-30. Low Profile Load Cell
(Courtesy, Futek Advanced Sensor Technology, www.futek.com)

A canister load cell is a tall cylindrically shaped transducer that has a flat base and hardened steel load application point at the top. The application point is usually a round dome with a convex radius, which helps to ensure that the load is applied through the sensitive axis of the load cell. Strain gauges are attached to an internal structure that is a machined block of steel mounted between the base and the load application point. This usually takes the form of a square or round column. The can that is placed around the column is often sealed so that the transducer can withstand water from rain or equipment washdowns. The canister load cell is one of the earliest load cell designs and can be made to have ranges up to the hundreds of thousands of pounds.

Beam-type load cells are simple machined steel bars with a round, square, or rectangular cross section. One end of the beam is attached to a support structure, while the other end supports the load to be measured. Strain gauges are attached to the top and bottom surfaces of the beam, which bends slightly as the load is applied. This bending puts the top gauge in tension and the bottom gauge in compression. Gauges mounted on the sides of the beam can be used for temperature compensation since they do not respond to the applied load.

In shear beam load cells (see Figure 4-31) an area of material has been removed from the beam in order to ensure that high shear forces occur near the reduced area. This is similar to an I-beam where the flanges resist beam bending, and the thin web between the flanges is an area of high shear loads. The shear beam load cell can be single ended or dual ended. On the single-ended load cell, one end of the beam is attached to a support structure while the other end supports the load. The dual-ended load cell is supported at both ends, and the load is applied to the center of the beam.

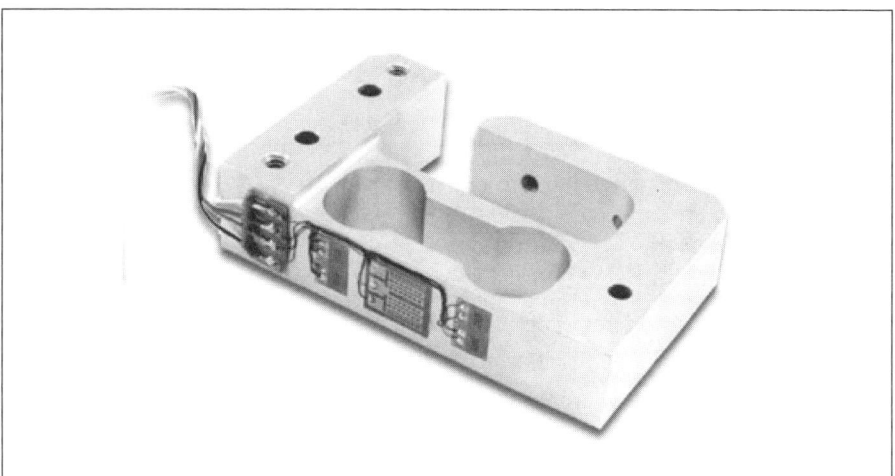

Figure 4-31. Shear Beam Load Cell
(Courtesy, Futek Advanced Sensor Technology, www.futek.com)

Pancake or low-profile load cells have a cylindrical-shaped body with small height to minimize the space required for installation. The flat bottom is attached to or supported by a smooth, flat surface, while the load is applied to the top of the transducer. The load is applied through a dome-shaped area or by a rod threaded into the center of the top surface.

4.7.3 Load Buttons

Load buttons are a special form of the low-profile load cell and are usually designed for low loads. They have a flat bottom surface and an integral dome-shaped extension on the top where the load is applied. Load buttons are usually very small transducers with ranges up to a few pounds.

The term *load button* can also apply to an accessory that is attached to a load cell in order to provide a dome-shaped load application point that directs the load through the transducer's sensitive axis.

4.7.4 Load Washers

Load washers are flat, circular load cells with a central hole. They are designed to be mounted under the head of a bolt in the same manner as a washer. These load cells measure the load applied by the bolt as it is threaded into a structure and tightened.

Table 4-7 summarizes the advantages and disadvantages of the various types of force transducers.

Table 4-7. Comparison of Load Sensors

Sensor Type	Advantages	Disadvantages
Canister	High upper range limit Allows load movement Can be hermetically sealed	Large size Side loads can be damaging Can be nonlinear
Bending beam	Inexpensive Easily made Good for low loads	Side load sensitive Off-axis load sensitive
Shear beam	Side load insensitive High accuracy	High-load, single-ended are expensive, difficult to mount
Pancake	Low profile Very low deflection	No load movement allowed

4.8 Strain Transducers

The two types of strain transducers discussed in this section are the resistive and the capacitive strain gauges.

4.8.1 Strain and Strain Measurement

Strain is defined as the change in length per unit length of an object when that object is subjected to a load. If the length of a 2-inch long steel bar decreases by 0.006 inches when loaded, then for a unit length of 1 inch the length of the bar decreases by 0.003 inches, and the strain is 0.003/1 = 0.003 inches per inch. This is often stated as 3000 micro inches per inch or 3000 micro-strains.

The applied load creates a stress in the object that is equal to the applied force divided by the unit area over which the force is applied. For elastic materials, the stress is linearly proportional to strain, and the material will return to its original length when the load is removed. Almost all materials have an elastic region where the stress is linearly proportional to strain. If the strain is too high, the material will be permanently deformed and will not return to its original length when the load is removed. At some higher level of strain the material will fracture. The purpose of applying strain gauges to test objects is to determine the level of stresses that the material is experiencing as a result of the forces acting on the object. These forces can be tensile or compressive, and they can be static or dynamic.

As we have seen in previous sections, the strain gauge is also used in many types of transducers, including pressure, velocity, acceleration, and force transducers.

4.8.2 Resistive Strain Gauges

The resistance strain gauge is based on the principle that the resistance of a metallic conductor will vary linearly with strain. When a wire is stretched (an increase in strain), its length will increase and its cross-sectional area will decrease. Both of these actions will increase the wire's resistance. Mathematically, the change in resistance divided by the original resistance is equal to a constant times the strain. The constant is known as the "gauge factor," which has a value of approximately two for commercial strain gauges. The original resistance of commercial strain gauges is commonly either 120 ohms or 350 ohms.

The strain gauge is constructed out of several parallel loops of wire that cover a certain area. The loops are long compared to the turns, which are very short and tight. The gauge is attached to the test object so the loops are parallel to the direction of strain. By using multiple loops of wire we have increased the length that is stretched and thus increased the change in resistance for a given strain. The short turns prevent any strains in the cross axis from significantly affecting the gauge output.

Strain gauges are now made from a metallic foil where the loops and turns are etched from the sheet of foil. An example of a foil strain gauge is shown in Figure 4-32. This allows the manufacturer to give the turns a much larger width than the loops. This minimizes the effect of cross-axis strains even more since the cross-sectional area and the length of the turns results in a very small resistance and a proportionally small change in resistance due to strain.

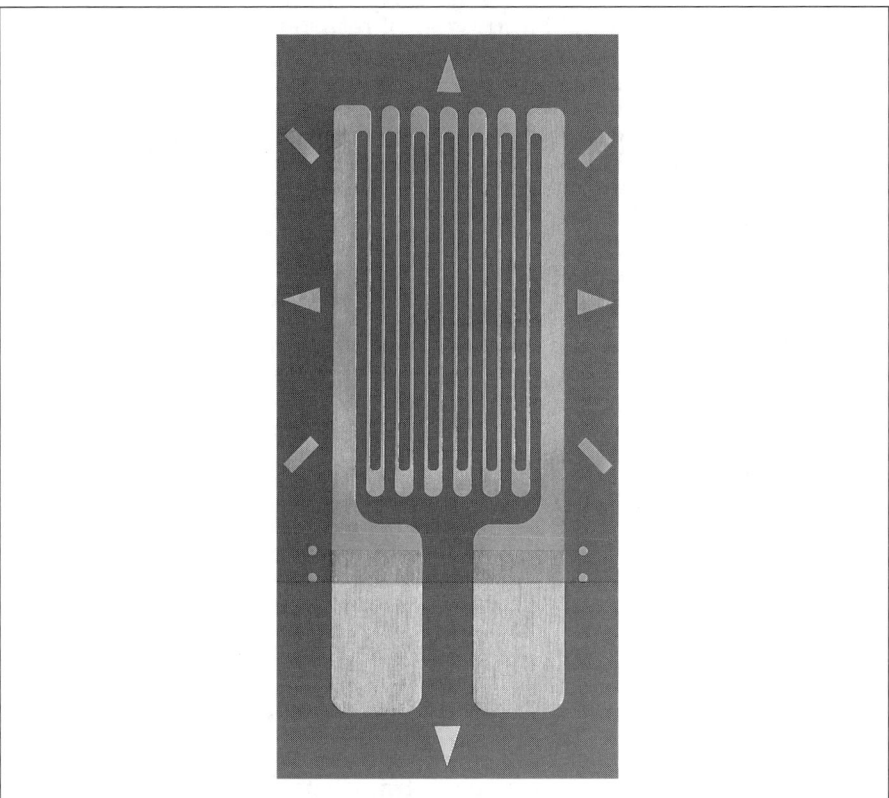

Figure 4-32. Foil Resistive Strain Gauge
(Courtesy, Vishay Micro-Measurements)

The foil is attached to an electrically insulated backing, which is bonded to the test object with special glues that harden to a very rigid material. This is required so that the strain of the test object is transferred by the glue and backing to the strain gauge loops. The complete strain gauge installation procedure consists of a number of steps. These include preparing the surface of the test object, measuring and marking the gauge location, cleaning the surface, gluing the gauge, curing the glue, testing the gauge resistance and insulation resistance, installing wiring, and waterproofing the gauge and wiring if necessary. The wiring must be done in a way that prevents the loads and movement of the wire and cable from being transferred to the strain gauge. This is often done by using intermediate solder tabs that are glued on the surface close to the gauge. The cable wires are soldered to these tabs, and a fine wire with a stress-relieving bend is used to connect the tab to the gauge.

Strain gauges are manufactured in a wide variety of shapes and sizes to suit specific applications. The length of the loop in typical gauges ranges from one-eighth of an inch to one-half of an inch. Since the gauge will respond to the average strain over its whole area, it is better to use the smallest gauge possible. Smaller gauges, on the other hand, are more susceptible to fatigue damage if the gauge is used to measure long-term dynamic strains. Gauges that have two or three strain gauges on the same backing are called rosettes. They are available with the gauges lying on the same plane or stacked on each other. Rosettes that have two gauges oriented at 90° to each other (see Figure 4-33) are used when the direction of principal strains is known and the magnitude is to be measured. Rosettes with three gauges mounted 45° or 60° apart (see Figure 4-34) are used to determine the direction of principal strains and the strains' magnitudes when the direction is unknown.

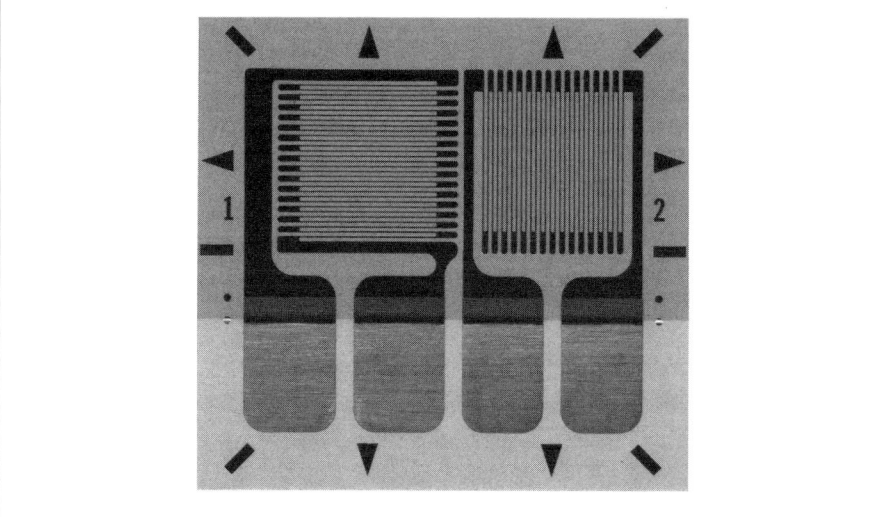

Figure 4-33. Ninety Degree Rosette
(Courtesy, Vishay Micro-Measurements)

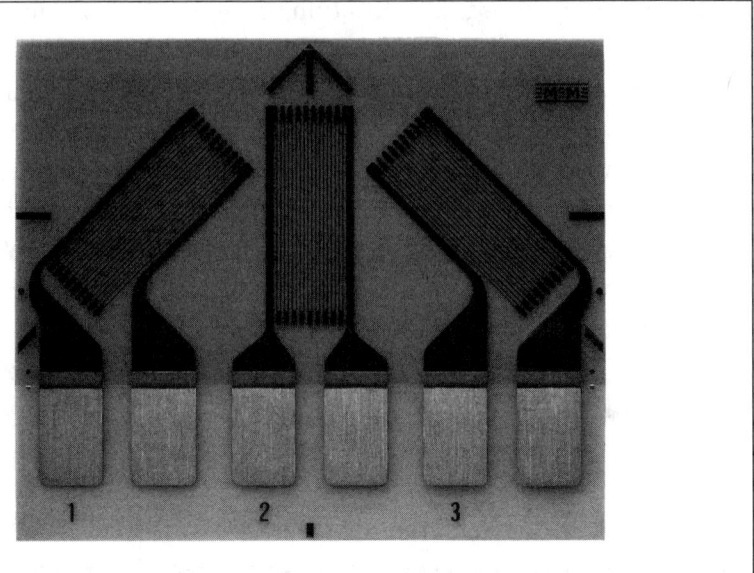

Figure 4-34. Forty-Five Degree Rosette
(Courtesy, Vishay Micro-Measurements)

Figure 4-35 illustrates a wide variety of strain gauges.

Figure 4-35. Foil Strain Gauge Assortment
(Courtesy, Vishay Micro-Measurements)

Figure 4-36 shows how strain gauges are typically packaged by the manufacturer.

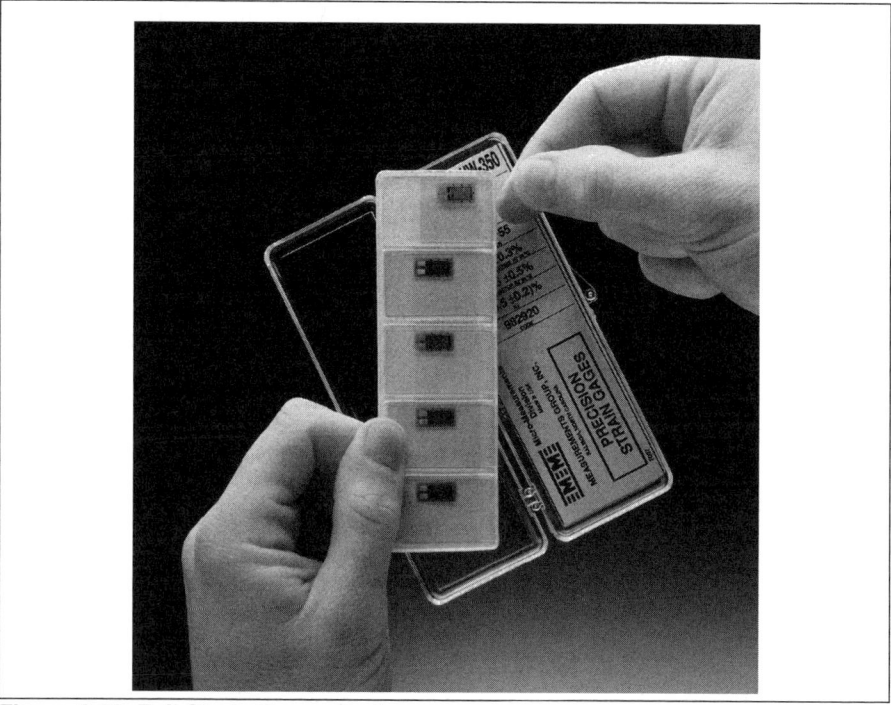

Figure 4-36. Foil Strain Gauge Package
(Courtesy, Vishay Micro-Measurements)

Temperature changes can affect the reading a strain gauge provides in two ways. The first is the fact that the gauge's resistance will vary with temperature. This is overcome by locating a second identical strain gauge in the same temperature environment as the test gauge but in a location that does not stress the gauge. The measurement circuit uses both gauges to subtract out the temperature effect and leave only the strain effect. The second error arises if the thermal expansion of the gauge material is not the same as the thermal expansion of the test object. This causes an apparent strain as the two parts expand different amounts due to temperature. To overcome this problem, strain gauge manufacturers make their gauges from alloys that have a variety of thermal expansion coefficients.

The strain gauge is connected into a Wheatstone bridge in order to measure the small change in resistance that occurs due to strain. The Wheatstone bridge commonly consists of four equal resistors connected in a diamond shape. A DC excitation voltage is applied to the top and bottom of the diamond while the output is measured across the sides of the diamond. When the resistances are equal, the output will be zero, and the bridge is said to be balanced. If one of the resistors changes value then the bridge becomes unbalanced and an output voltage will appear. In the simplest circuit, the strain gauge becomes one of the resistors in the bridge. This one active arm bridge is known as a "quarter bridge circuit." A half bridge circuit uses two gauges that experience equal but opposite strains. Locating these gauges in adjacent arms of the bridge causes the output to double compared to the output of the quarter bridge circuit. A full bridge uses two pairs of gauges that experience equal but opposite strains. The output of a full bridge is twice that of a half bridge and four times that of a quarter bridge.

4.8.3 Capacitive Strain Gauges

Capacitive strain gauges are specifically designed to measure large strains at high temperatures, up to 1500°F. The gauge is an assembly of two cylindrical capacitor plates that slide within a concentric outer plate that is common to the other two. The two capacitors are connected as adjacent arms in a bridge circuit, and when the two cylindrical plates are centered within the outer plate, the bridge is balanced and the output is zero. When the two plates move they cause a change in plate area with respect to the common plate. One area will increase while the other will decrease. A corresponding change in capacitances will unbalance the bridge and result in an output that is proportional to the displacement caused by the strain. These units are commonly spot welded into place on the test object and used for long-term monitoring of the strain. Figures 4-37, 4-38, and 4-39 illustrate the construction of these capacitive strain gauges.

Chapter 4 – Transducer Types   115

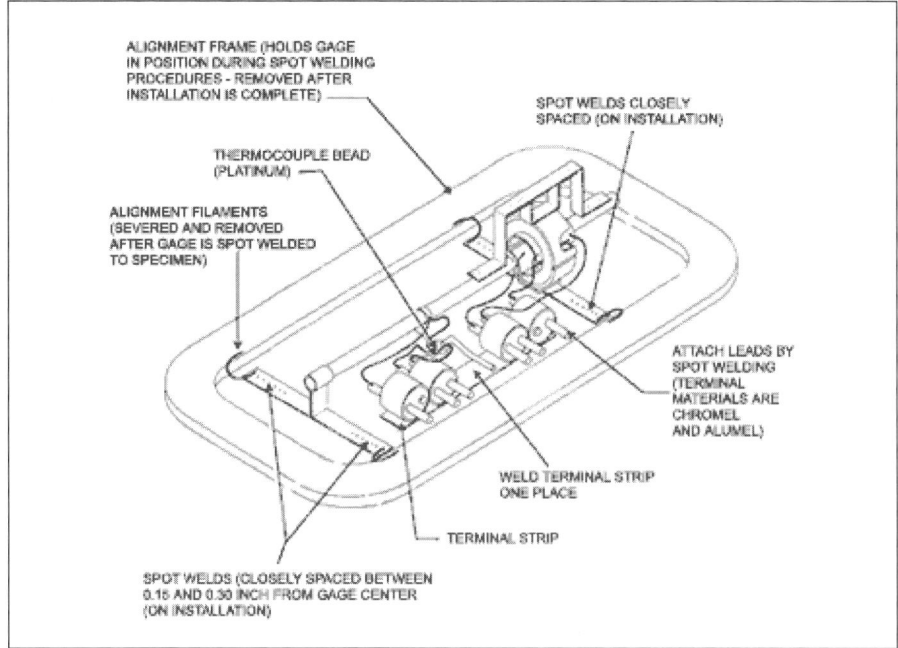

Figure 4-37. Capacitive Strain Gauge Design
(Courtesy, Hitec Products, Inc.)

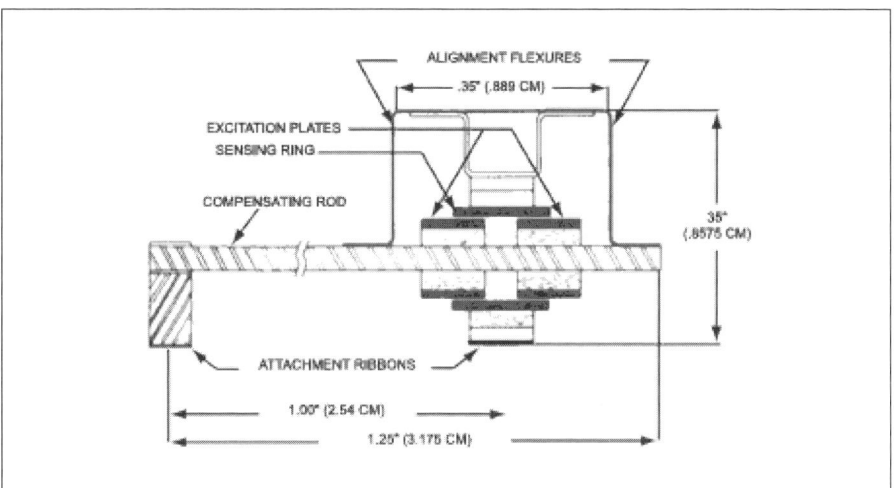

Figure 4-38. Capacitive Strain Gauge Side View
(Courtesy, Hitec Products, Inc.)

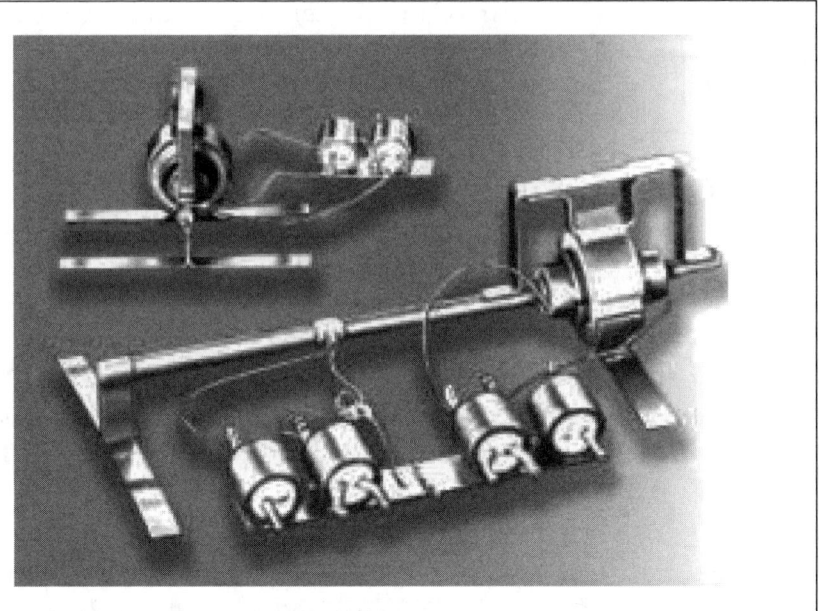

Figure 4-39. Capacitive Strain Gauge Details
(Courtesy, Hitec Products, Inc.)

5
Signal Conditioning

In a typical measurement system, signal conditioning is the equipment that is located between the transducer and the display. As its name suggests, this equipment conditions the transducer signal into a form that is usable by the display.

In some of the transducers described in chapter 4, the signal conditioning electronics is built into the body of the transducer. This type of transducer has the advantage of providing a high-level, low-impedance output signal that is relatively immune to noise contamination caused by radio frequency interference (RFI) or electromagnetic interference (EMI). The disadvantages of this type of transducer are the limited temperature environment, which is a result of the electronic components and the larger physical size of the transducer.

Other transducers have some or all of the signal conditioning built into an in-line unit that is located near the transducer but away from any extreme test environment. This is a compromise solution that may be advantageous in some test situations. Its disadvantages are that both the number and the adjustability of the in-line unit's signal conditioning capabilities are usually limited, and the unit may be located in a less than ideal place as far as accessibility and environment are concerned.

The most common situation is to have the transducer mounted directly on or very close to the test object. Cable runs of tens to hundreds of feet are used to carry the transducer's output signal to the signal conditioning equipment, which is located in a control room that also houses the display equipment and the human system operators. For some transducers, the cable also carries the excitation voltage that is required by the transducer and provided by the signal conditioner. The main advantage of this type of system is its flexibility. The typical bench-mounted or rack-mounted signal conditioner used in this type of system has adjustments that can be tailored to suit the transducer, the display, and the test requirements. A strain gauge amplifier can have adjustable excitation voltage, adjustable amplification, adjustable zero setting, built-in bridge completion resistors, adjustable filter settings, and multiple outputs. This type of signal conditioning can be used with strain gauge transducers from different manufacturers and with almost any type of display. The main disadvantage of this type of system is that long cable runs are required between the transducer and signal conditioner, and these are susceptible to noise pickup.

The most common electrical output from test measurement signal conditioning equipment is a DC voltage range that is linearly proportional to the transducer's input range. A typical DC voltage range is 0 to 10 volts, where 0 volts is the output from the signal conditioner when the transducer input is zero, and 10 volts is the output when the transducer input is at full scale.

This DC voltage can be connected to a wide variety of display equipment, which can be scaled or programmed to convert the DC voltage level into an equivalent number of input units that the transducer is sensing. A strain gauge pressure transducer with an input range of 0 to 1000 psig may have a full-scale output sensitivity that is specified as 2 millivolts per volt. This means that the output from the transducer bridge at an input of 1000 psig will be 2 millivolts for every volt of bridge excitation voltage. If the bridge excitation voltage is 10 volts then the output from the transducer bridge will be 20 millivolts at an input of 1000 psig. At 0 psig, the bridge will be balanced, and the output will be 0 millivolts. Since the transducer is linear, the bridge output will be 10 millivolts at an input pressure of 500 psig. To provide the 0-to-10-volt signal, the signal conditioning equipment must include an amplifier that has a gain of 500. At a full-scale input of 1000 psig the amplifier will multiply the 20 millivolt bridge output signal by 500 to produce the required signal conditioner output of 10 volts. An input pressure of 500 psig produces a bridge output of 10 millivolts, which will be amplified by a factor of 500 to produce an output signal of 5 volts. An input of 0 psig will produce an output of 0 volts.

In this example, the signal conditioning equipment consists of a DC power supply that has a stable, regulated output of 10 volts and a voltage amplifier with a gain of 500. In practice, these units would be packaged into a single module, along with adjustments for amplifier gain, amplifier zero, and input and output connectors for cable connections to the transducer and display. The adjustable amplifier gain and zero are provided to calibrate the system so that 0 psig produces exactly 0 volts and 1000 psig produces exactly 10 volts. This is required since no transducer has an output that is completely ideal. The strain gauge bridge may not be completely balanced at 0 psig, and the output at 1000 psig may not be exactly 20 millivolts due to physical variations in the construction of the transducer and variations in the strain gauges and their installation. This voltage amplifier is one of the simplest forms of signal conditioning, and it is likely to be found as a building block within more complex signal conditioners. It is easy to see how a few integrated circuit chips and a few discrete components can be added to the transducer housing to make a transducer with built-in signal conditioning.

5.1 Voltage Amplifiers

The voltage amplifier is the most common form of signal conditioning equipment whether it is being used on its own or as part of a more complex system. Almost all sensor signals start out as a low-level electrical signal of some type. They may be low-level DC millivolt signals from strain gauge bridges or thermocouples or small capacitive or AC millivolt signals from other transducers. In the case of the latter transducers, some electronic circuitry is then required to transform the original signal into a DC millivolt signal. The DC millivolt signals are then input into a voltage amplifier in order to produce a higher-level DC voltage signal that can be interfaced to the display equipment.

The requirements of a high-quality voltage amplifier include:

- High input impedance
- Low output impedance
- Stability
- Low noise
- High gain
- High bandwidth

The high input impedance is required so the amplifier does not load the transducer output to the point that the voltage seen at the input of the amplifier is less than the transducer output would have been if the amplifier were not present. The transducer can be modeled as an ideal voltage source in series with a resistor that has a value equal to the transducer's output impedance. The voltage level of the ideal source is proportional to the value of the parameter being measured by the transducer. The ideal voltage source maintains a constant voltage output regardless of the current that is drawn. The output impedance resistor is connected to another resistor that is equal to the amplifier's input impedance. This input impedance resistor is the value that appears across the amplifier's input terminals. Our model is now that of a voltage divider that has the transducer's ideal voltage source supplying the current through the two series-connected resistors and the voltage amplifier measuring the voltage caused by the current through the input impedance resistor.

In Figure 5-1,

$$V_A = V_T (R_I / (R_I + R_O))$$

From this equation we can see that if the input impedance of the amplifier (R_I) is equal to the output impedance of the transducer (R_O) then the voltage measured by the amplifier is only one-half the ideal voltage of the transducer. If the input impedance is 1,000 times that of the output impedance then the voltage measured by the amplifier is 99.9% of the transducer voltage. If the input impedance is 1,000,000 times that of the output impedance then the voltage measured by the amplifier is 99.9999% of the transducer voltage. This illustrates the need for voltage amplifiers that have high input impedance. Values of one megohm are typical, and many amplifiers have input impedances of 100 megohms.

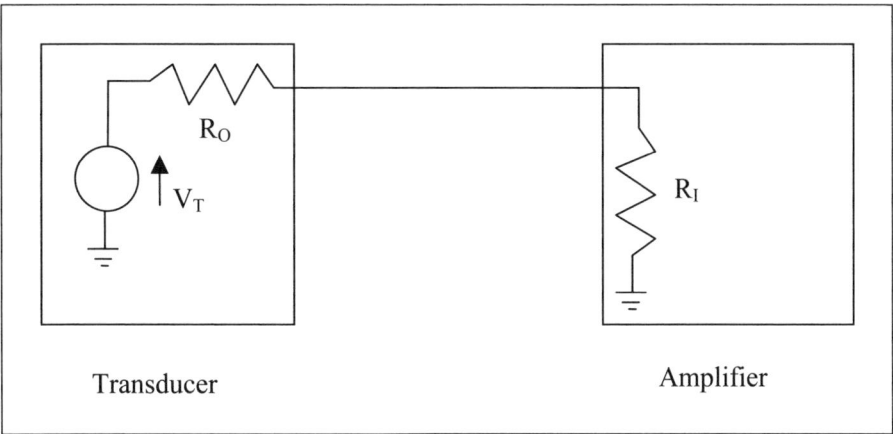

Figure 5-1. Transducer and Amplifier Model

The requirement that the amplifier have low output impedance can be explained using the example just given. The display equipment that is connected to the amplifier output must also have high input impedance compared to the output impedance of the amplifier. Keeping the amplifier's output impedance low minimizes the need for the display equipment to have excessively high input impedance.

Stability of all parts of the measurement system is an obvious requirement. To maintain the system stability, the voltage amplifier must provide a constant zero and a constant gain. The zero and gain must be constant with time and with variations in ambient temperature. This is accomplished in the amplifier design by selecting components that have low drift and low or balanced temperature coefficients.

The electronic components that are used in amplifiers contribute noise to the signal. Integrated circuit operational amplifiers will generate some noise that can get added to the measurement signal. Resistors generate noise depending on their value, the signal bandwidth, and the ambient temperature. Other noise can be caused by EMI and RFI emissions. The amplifier designer has to employ correct circuit-board design and shielding techniques to reduce the coupling of this EMI and RFI emissions noise into the measurement signal. Noise generated at the front end of the amplifier is troublesome since it will be amplified along with the measurement signal.

High amplifier gain is required to boost the low-level transducer signal to the high levels required by the system's display section. The strain gauge pressure transducer example described earlier in section 5.0 required a gain of 500. Lower level signals from sensors such as thermocouples will require gains several times larger.

High bandwidth describes the amplifier's ability to provide a constant gain over a wide frequency range. This high bandwidth is required when carrying out ballistic testing, shock or impact testing, and sound testing where the measured phenomena are changing at high speed. The integrated circuit operational amplifiers that are used to construct voltage amplifiers have a specification called the "gain-bandwidth product." This is a limit to the gain multiplied by the bandwidth, and it causes a tradeoff between the two specifications. If high bandwidth is required then more than one stage of amplification may be required to also produce the high gain. This results in more components and more complex circuit and circuit board design.

5.2 Linearizing Amplifiers

Linearizing amplifiers are voltage amplifiers that have additional circuitry to convert a nonlinear transducer signal into a linear amplifier output signal. Sensors such as thermocouples and RTDs are examples of transducers that have an output signal that is not linearly proportional to the input being measured. The linearizing function can take place in the signal conditioning equipment or in the display equipment. If the display equipment is computer based then it is usually easier to carry out the linearization using software. The exception would be if the processing time for linearization became too great and impacted the data acquisition speed. Signal conditioning–based linearization is required for simple display equipment or for equipment like signal analyzers that requires a linear signal input.

5.3 Strain Gauge Amplifiers

Strain gauge amplifiers are voltage amplifiers with added features that allow the amplifier to interface with strain gauges or strain gauge–based transducers. These features include:

- Bridge excitation power supply
- Bridge completion resistors
- Adjustable gain
- Adjustable zero
- Adjustable filters
- Multiple outputs
- Bridge balance adjustment
- Pushbutton shunt calibration

Typical strain gauge amplifiers are shown in Figures 5-2 and 5-3.

Figure 5-2. Individual Strain Gauge Amplifier
(Courtesy, Vishay Micro-Measurements)

Figure 5-3. Rack-Mounted Strain Gauge Amplifiers
(Courtesy, Vishay Micro-Measurements)

The strain gauge or transducer requires a stable DC voltage power supply to provide the excitation voltage to the Wheatstone bridge circuit. The built-in power supply is often adjustable over a 1-to-15-volt range so as to accommodate bridges that have different input impedances. Higher excitation voltage results in higher bridge output but also results in more current through the strain gauges and thus more self-heating and temperature-induced errors. A 120-ohm bridge typically uses a 5 volt excitation voltage, while a 350-ohm bridge uses a 10 volt excitation voltage. Some amplifiers have the ability to sense the excitation voltage at the transducer by running an extra pair of wires from the amplifier. The power supply circuit reads this voltage and adjusts the output to provide the exact voltage regardless of any voltage drop along the length of the wires.

When only one or two strain gauges are used to make the measurement, the other resistors that are required to complete the four-arm Wheatstone bridge can be located within the amplifier housing. These can be soldered in as required or permanently connected and switched in as required.

The adjustable gain is needed to provide the required output voltage for a given input signal. Often, a series of pushbutton selectable gains are provided along with an adjustable trim potentiometer to fine-tune the gain.

An adjustable zero or amplifier balance potentiometer is provided to cancel any offset voltages present in the amplifier. This is often done by removing the transducer excitation voltage with a pushbutton and adjusting the amplifier output to zero volts.

Adjustable filters are provided to limit the frequency bandwidth of the amplifier's output voltage. In a strain gauge amplifier these are usually low-pass filters that do not affect the DC and low-frequency content of the signal but block high-frequency signals that may not be of interest. High-frequency noise is also blocked by the filters. A series of pushbuttons is provided to select the required upper frequency limit that will be passed to the amplifier output.

Multiple outputs may be provided to suit different display equipment. These may include a high-voltage output for meters, recorders, and data acquisition systems; an output that is compatible with magnetic tape recorders; and a high-current output to drive a galvanometer style recorder.

A bridge balance adjustment is used to trim the bridge resistors until there is zero bridge output for zero measurement signal at the transducer input. Some amplifiers have a battery-powered circuit that remembers the balance setting even if the amplifier is shut off. This is useful in test conditions where the process cannot easily be shut down in order to rebalance the transducers.

Shunt calibration is a technique that places a high value resistor in parallel with one of the arms of a Wheatstone bridge while the bridge is balanced. This effectively lowers the resistance value of that arm and unbalances the bridge so that an output is created. The value of resistance that causes a certain strain can be calculated. For strain gauge transducers, however, the usual procedure is to calibrate the transducer and amplifier using a physical standard and then select a resistor that produces some percentage (80%) of the amplifier full-scale output. This resistor is installed in the amplifier and connected to the bridge with a pushbutton when required. This technique only checks the calibration of the amplifier and does not replace a complete system calibration. A system calibration of a pressure transducer and amplifier would require that a series of known pressures be applied over the full range of the transducer and the amplifier output be noted at each step. This would stress the transducer diaphragm and change the resistance of the strain gauges. The shunt calibration does not stress the transducer diaphragm and thus does not calibrate the transducer. Transducer manufacturers will often provide a resistor value or even a resistor that will create a specified output based on the initial transducer calibration.

5.4 Charge Amplifiers

Charge amplifiers (see Figure 5-4) are designed to interface with the high-output impedance of the piezoelectric transducer and convert the charge output into a low-output impedance voltage signal. The simplest form of the charge amplifier is an inverting operational amplifier that has a capacitor to transfer the charge from the transducer crystal to the amplifier input and a feedback capacitor to set the amplifier gain.

Figure 5-4. Charge Amplifier
(Courtesy, Brüel & Kjær)

The piezoelectric transducer will have a charge output that is linearly proportional to the input. A pressure transducer's output could be 10 picocoulombs per psi, while an accelerometer could have an output of 10 picocoulombs per g.

The output from the amplifier that has a feedback capacitor value of 100 picofarads is:

$$V_{out} = \frac{10 \text{ picocoulombs / psi}}{100 \text{ picofarads}} = 100 \text{ millivolts / psi}$$

This relationship is not affected by the transducer's or cable's capacitance, and changes in cables or cable length can be made without affecting the system's calibration.

A large value resistor is placed in parallel with the feedback capacitor, and the product of the resistor and capacitor values sets the time constant of the system. This time constant determines the low-frequency cutoff of the amplifier. The value of the feedback resistor is typically in the giga-ohm range.

The materials and components used in the amplifier must have a very high insulation value so the signal charge does not dissipate. This includes connectors, cables, electronic components, and circuit board materials. Even low leakage currents can saturate the amplifier because of the high-value feedback resistor. The amplifiers must be kept free of any dirt or moisture that could lower the insulation value or provide a leakage path.

The features that are included in a typical laboratory charge amplifier include:

- Selectable high-pass filter setting
- Selectable low-pass filter setting
- Integrator circuits to convert acceleration into velocity and velocity into displacement
- Adjustable gain or selectable transducer sensitivity matching

5.5 Filters

Filters are used to remove signals that have a specific frequency or frequency range from the overall measurement signal. This is done to remove noise, simplify data analysis, improve signal display, and limit undesirable transducer characteristics. While filters are often built into the signal conditioning equipment, such as strain gauge or charge amplifiers, they can also be part of the transducer or the display equipment. In the case of the transducer, any dampening that slows down the motion of physical components can be considered a form of filtering. The majority of filters are electronic circuits that are built into one or more of the measurement system components. In some cases, the filter is constructed as a stand-alone piece of equipment with multiple channels and adjustments of the frequency cutoff point.

The most common form of filter is the low-pass filter. It allows low-frequency signals to pass through it but reduces or attenuates higher frequency signals that are above a specific cutoff or corner frequency. The high-pass filter does not affect high-frequency signals but attenuates low-frequency signals below a specific corner frequency. The band pass filter is a combination of the low-pass and high-pass filters. It passes a certain range of frequencies but attenuates low-frequency signals below a specific corner frequency and attenuates high-frequency signals above a specific corner frequency. The band reject or notch filter is the opposite of the band pass filter. It attenuates a narrow range of frequencies but allows the passage of higher and lower frequencies.

The ideal filter would not affect the signal until the cutoff frequency is reached. At this point the filter would immediately reduce the signal to zero. In practice, the electronic filter reduces the signal on a more gradual basis once the cutoff frequency is reached. The rate at which the reduction takes place is called the "filter rolloff" and is determined by the number of filter sections that are cascaded together. The rolloff or slope is described by the attenuation that occurs per decade change in frequency. The attenuation has units of decibels (db).

$$\text{Attenuation (decibels)} = 20 \log_{10} (\text{filter output} / \text{filter input})$$

The single stage or single pole or first-order filter has an attenuation equal to 20 db per decade or 20 db per tenfold increase in frequency. 20 db represents a filter output–to–filter input ratio of 0.1, so another way of describing the rolloff or slope is to say that the output is reduced to one tenth of the input for every tenfold increase in frequency.

A two-pole or second-order filter has an attenuation equal to 40 db per decade, which is the same as saying that the output is reduced to one hundredth of the input for every tenfold increase in frequency.

The corner frequency is the frequency at which the attenuation is equal to 3 db.

The filter can also cause time delays or phase shifts in the signal. This can lead to signal analysis problems when signals that are filtered in different ways are to be compared. Signal distortion can also be caused by the nonlinear phase response of some filters.

5.6 Signal Isolators

Signal isolators (see Figure 5-5) are used to create a physical separation in the wiring between components in a measurement system. They form a wireless transmission system that takes an electrical signal as an input, converts it into a signal that can be transmitted across an air gap, and then reconverts the signal back into an equivalent electrical signal at the output. Two common methods of signal transmission are transformer coupling and optical coupling. Transformer coupling uses a transformer's primary and secondary to isolate and couple the input to the output. Optical coupling uses an electronic light transmitter such as a photodiode to transmit the input signal to a light receiver such as a phototransistor. Figure 5-6 shows the block diagram of a typical signal isolator.

The most common use of isolators is to eliminate ground loops that occur when multiple pieces of equipment are grounded at different locations that have different ground potentials. This causes current to flow in the grounded conductors, which can lead to error signals being added to or subtracted from the measurement signal. These error signals can potentially be large enough to damage the input circuits of signal conditioning or display equipment.

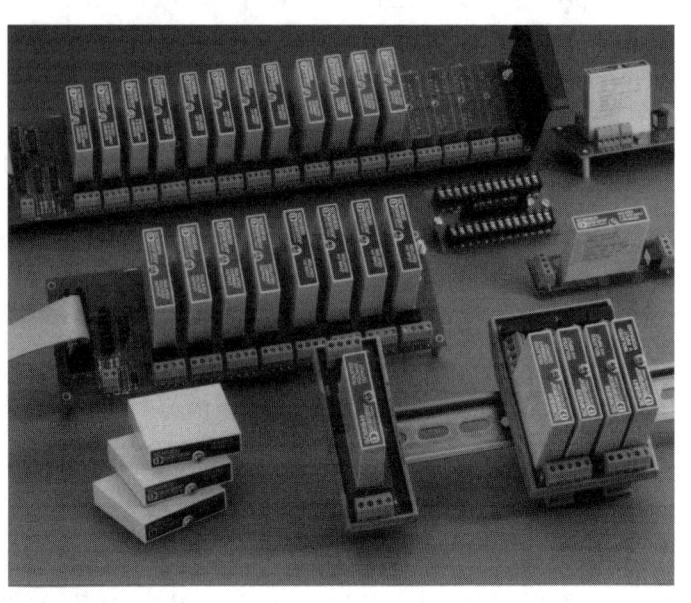

Figure 5-5. Signal Isolators
(Courtesy, Dataforth Corporation)

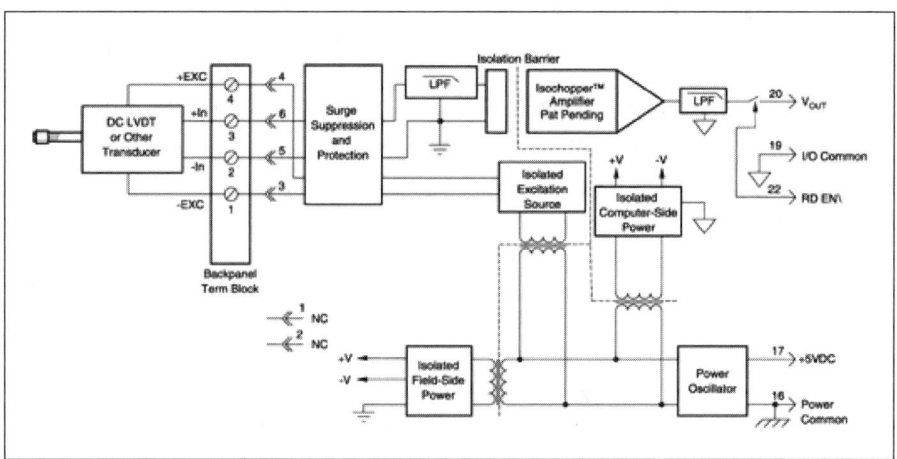

Figure 5-6. Signal Isolator Block Diagram
(Courtesy, Dataforth Corporation)

5.7 Analog-to-Digital Converters

Analog-to-digital converters transform the analog voltage signal from a transducer or a signal conditioner into an equivalent digital signal that can be read by a digital display or computer-based data acquisition system. The analog-to-digital converter (ADC) is usually built into the digital display or data acquisition system and rarely exists as a separate part of a measurement system.

The primary specification of an ADC is its resolution, which is determined by the number of bits present in the digital output.

$$\text{Resolution} = 1 / 2^n \text{ where n equals the number of bits}$$

An 8-bit ADC will have a resolution of $1 / 2^8 = 1 / 256$.

A 12-bit ADC will have a resolution of $1 / 2^{12} = 1 / 4096$.

A 16-bit ADC will have a resolution of $1 / 2^{16} = 1 / 65536$.

If the input voltage range to the ADC is 0 to 10 volts then:

- The resolution of an 8-bit ADC is 10/256, which is approximately 40 millivolts.
- The resolution of a 12-bit ADC is 10/4096, which is approximately 2.5 millivolts.
- The resolution of a 16-bit ADC is 10/65536, which is approximately 0.15 millivolts.

The other important ADC specification is conversion rate, which indicates the time taken to convert a sample of the analog voltage into a digital equivalent signal. This is primarily based on the type of electronic circuit used to perform the conversion. The dual slope integration technique has a slow conversion rate and is used in digital meters. The successive approximation technique is capable of hundreds of thousands of conversions per second, while the flash technique is capable of millions of conversions per second.

5.8 Differential versus Single-Ended Inputs

Amplifiers and other signal conditioning units have inputs that are described as single ended or differential.

The single-ended input arrangement shown in Figure 5-7 has an input signal connection and a reference connection that is usually ground. The output signal (V out) from this circuit is the input signal (V in) multiplied by the amplifier gain (A) and is with respect to the reference input (gnd). This type of input is used when the output of the transducer or equipment connected to the input has the same ground as the amplifier.

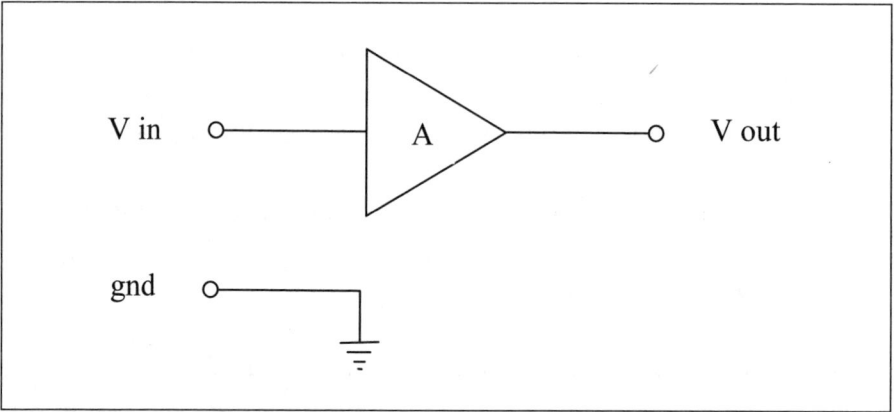

Figure 5-7. Single-Ended Input

If the grounds are at different potentials then ground loop currents will flow, and these currents will distort the signal. Noise that is coupled to the signal wiring will be amplified along with the input signal so single-ended connections are limited to high level (greater than 100 millivolts) inputs that do not require high gains.

Single-ended inputs cannot be used on balanced output transducers such as strain gauge bridges where the bridge outputs are both at some voltage above ground and it is the difference in voltage caused by bridge unbalance that is to be measured. Grounded thermocouple outputs cannot be connected to single-ended inputs since the grounded reference input will basically short-circuit the signal source.

Single-ended inputs are mostly used for any units that are located after the first piece of signal conditioning equipment. These units are usually located close together, have the same ground potential, and provide a high-level signal that is referenced to the ground.

Figure 5-8 shows the arrangement for a differential input signal conditioner that has a positive input (V in +), a negative input (V in −), and a reference input that is usually ground. The output (V out) is equal to the difference between the positive and negative inputs multiplied by the gain (A).

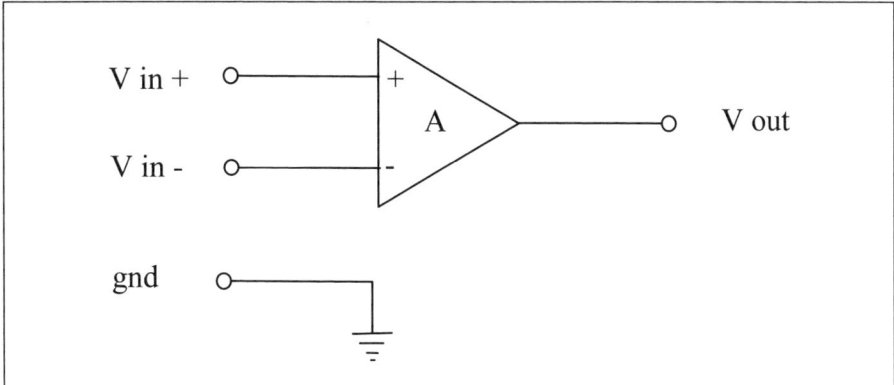

Figure 5-8. Differential Input

This type of circuit can be used when the ground potentials of the transducer and signal conditioner are different and where the transducer output is balanced, as in the case of a strain gauge bridge. The two output points of a bridge are at a potential that is a few volts above ground. The bridge unbalance causes a difference between the two points that is only a few millivolts. The differential input subtracts out the common voltage signal and amplifies the millivolt level difference signal that is proportional to the transducer input signal.

The differential input circuit also minimizes noise that is coupled to the two input connections. This noise will likely have the same level on both leads and will be subtracted out by the differential input. If a single-ended transducer is used with a differential input signal conditioner, then the unused input should be connected to the ground reference so that noise is minimized.

In summary, the differential input should be used where different ground potentials exist, for long cable lengths, for low-level signals, and for balanced output transducers.

6
Transducer Installation

Installing the transducer correctly is as important as selecting and integrating it correctly with the balance of the signal conditioning system. Ideally, the installation should not affect the test environment, and the test environment should not affect the transducer. The test environment parameters must be known before selecting the transducer, and the effects of the parameters on the transducer's performance must be evaluated and accounted for in the data reduction phase. The transducer's specification sheets should specify limits to the test environment parameters and detail how much the transducer output will be affected for a given change in the environment. These parameters include temperature, pressure, moisture, radiation, vibration, shock, and corrosive atmosphere. Moisture specifications can include relative humidity, water spray, or depth of water submersion.

The user's manual or specification sheet should also outline recommendations for installing the transducer. These may include recommendations involving mounting bolt torque, mating thread torque, surface flatness and finish, mounting dimensions, method of sealing, sealing torque, off-axis load limit, minimum bend radius, lubrication requirements, and tests required to validate installation. Other specifications include electrical connector requirements, wiring information, excitation voltage and current requirements, and output signal load limit. Wiring information will include connector pin designations, wire color codes, wire size, maximum wire length, cable type (co-axial, twisted pair, etc.), and shielding requirements.

If the manufacturer cannot supply certain specifications, then the user must carry out tests to determine the effect of any missing environmental or installation parameters that he or she may encounter while using the transducer. The user should be aware that the specifications for parameter effects are usually given independently of each other and do not take into account cases where multiple parameters are acting at the same time. User tests may again have to be carried out to determine how the transducer will operate under actual test conditions. These types of tests are time consuming and expensive, and it may be more productive to try and limit one or more of the parameters. Cooling a transducer with an air flow or water cooling jacket is an example of this approach. Adding shielding to minimize radiation is another example.

In most cases, adding a transducer to the test environment changes the environment. The user must evaluate whether the addition will impact the measurement being made and whether it will change how the item being tested operates. Some things to consider include the transducer's mass changing the vibration characteristics, the transducer's size and position changing flow profiles, the transducer's installation causing heat conduction, the installation of the transducer causing changes in test item stiffness, and the mounting method affecting the transducer's frequency response.

6.1 Wiring and Cabling

The wiring and cabling used to connect the transducer to the signal conditioner are an important part of the measurement system. They must transmit the transducer signal without change or limit, and they must prevent external influences from affecting the signal. Properties such as wire resistance and cable capacitance can influence the signal, while cable shielding can reduce the effects of electrical noise that may contaminate the signal.

Almost all electrical equipment found in a laboratory or plant has the capability of inducing electrical noise into a measurement circuit. Rotating equipment such as motors and electrical generators are prime causes of noise, but stationary sources such as AC power cables and fluorescent lights can also cause considerable noise. The noise level is generally higher if the source voltage or current are high or if the source is close to the measurement system and its wiring.

The noise source can be an electrical field or a magnetic field. The varying charge that is produced by a changing electrical field can be capacitively coupled to the signal wiring, while the changing currents in equipment or wiring can cause varying magnetic fields that inductively couple with the signal wiring and cause a varying current to flow.

Capacitive coupling to the signal wiring can be minimized or eliminated by using a shield. A shield is a grounded metallic covering that surrounds the signal wiring and acts as an equipotential termination for the electrical fields before they can be capacitively coupled to the signal wiring. The shield must provide a low impedance path to ground so the charge is effectively dissipated. The shield should be grounded at one point only to prevent ground loops.

Inductive coupling can be minimized by using twisted pairs of signal wires. Tightly twisting the wires will pull them close together and reduce the wire loop area within the magnetic field. This will reduce the level of the induced voltage. Twisting also minimizes the electrical field effect since short equal lengths of wire are alternatively placed in two different positions relative to the electrical field. This tends to cancel out the coupled voltages in adjoining lengths of wire.

Since most plant equipment will produce both electrical and magnetic fields it is usual to combine twisted-pair wiring with shields. Cables are also available that have individually shielded pairs of wire. These are useful for transducers such as strain gauge transducers that require one pair to carry the high-level excitation voltage and the other pair to carry the low-level strain gauge signal. This prevents any noise in the excitation voltage from being coupled to the strain gauge signal. Cable shielding can be made out of a braided mesh of fine copper wire or out of a metal foil and enclosed drain wire. The braided shield provides approximately 95% coverage due to the spaces within the mesh, but it does provide a very low resistance path. The foil shield provides 100% coverage but has a somewhat higher resistance path to ground.

6.2 Grounding

Cable shields must be correctly grounded to eliminate or reduce the chance of electrical noise getting coupled to the transducer signal. The shield must always be grounded for it to be effective, and it should only be grounded at one end so that ground loop currents do not flow in the shield. If the transducer is supplied with a mating connector and cable, check to determine if the shield is already connected to ground within the connector. In this case, do not connect the other end of the shield to ground. In the ideal measurement system, all the grounds should be terminated at one point. This may require that similar grounds are terminated on low resistance bus bars and then all the bus bars are subsequently connected to a single ground.

6.3 Connecting to the Test Rig

Transducers have to be mounted on the test rig in a way that allows them to function completely and accurately while preserving the safety of personnel and equipment and minimizing their effect on the measurement and the test environment.

Allowing the transducer to function completely and accurately means installing the transducer so its intended motion is not restricted, its measurement accuracy or response time is not limited, and the sensing element's access to the test environment being measured is not restricted. Load cells, for example, have to deflect a small amount in order to provide a measurement signal. This motion should not therefore be restricted by the method used to mount the transducer, such as threaded studs being engaged too deeply and making contact with the transducer body. Accelerometers must have a mounting that completely transmits the test object's motion to the body of the accelerometer. A mounting that is too loose or misaligned will cause measurement inaccuracies. A mounting that is too soft or compliant will not transfer higher frequency motions to the accelerometer. Temperature sensors such as RTDs must be immersed a certain depth into the fluid so the entire resistance element is exposed to the fluid.

Safety issues should always be considered when installing and operating test instrumentation. This is extremely important when the test environment is at high pressure or temperature or is hazardous because of its chemical or radiation properties. Transducers must be selected so they match their intended environment and have sufficient overload capabilities to protect against out-of-range events. Safe access to the transducer must be provided if the operator has to make adjustments during its use.

Adding a transducer almost always affects the test environment to some degree. The over-long stem of a temperature transducer will conduct heat either into or out of the test environment. This effect can be minimized by insulating the stem to prevent heat transfer to or from the atmosphere. Adding an accelerometer can change the test object's frequency characteristics either because of the additional mass or the stiffening caused by the mounting method. In some cases, even the connecting cables can influence the vibration measurement in the same ways.

As mentioned previously, other considerations affecting the installation of transducers include mounting bolt torque, mating thread torque, surface flatness and finish, mounting dimensions, method of sealing, sealing torque, off-axis load limit, minimum bend radius, lubrication requirements, and tests required to validate installation.

The torque requirements for mounting bolts will usually be specified by the transducer's manufacturer. These mounting bolts secure the transducer to the test rig or to an adapter that is attached to the rig. The torque requirements may guarantee a secure vibration-resistant mounting, a certain mounting stiffness, or a leak-free seal. The seal may be accomplished by compressing a gasket or o-ring by applying a specified bolt torque. The order of bolt tightening as well as the level of final torque may be specified in the requirements.

Mating threads such as pipe threads are often used to mount pressure transducers to test rigs. These connections rely on the wedging action of a tapered thread to provide the pressure seal, and the required torque may be specified by the manufacturer. In other cases, the seal is accomplished using parallel or straight threads and a seal ring that is compressed by applying a specified torque to the threaded portion of the transducer.

Surface flatness may be required so the transducer's body is not strained or deformed when it is mounted to the surface of the test rig. This is critical when mounting low-profile load cells or accelerometers to a surface. A certain surface finish may be required to guarantee a pressure seal, allow a strain gauge to bond correctly, or allow good coupling between an ultrasonic transducer and the surface.

Mounting dimensions for the transducer are usually provided by the manufacturer. These can include the spacing of mounting bolt holes, the minimum spacing between transducers, the counter-bore dimensions around faces of inductive and capacitive probes, stand-off distance for laser displacement probes, and the dimensions of the mating threads.

All load cells are designed so that the applied load passes through a specific or primary axis. Off-axis loads that are applied at an angle to the primary axis can be resolved into two components. One component will pass through the primary axis and be measured by the load cell. The other component will act at right angles to the primary axis and is called the "side load." The manufacturer will specify a maximum side load that the load cell will withstand and also the degree to which the load cell output is affected by a given side load. Load cell adapters are available to minimize or eliminate off-axis loads. These include load buttons and ball bearings for compression loads and clevis assemblies and self-aligning rod end bearings for tension loads.

The mineral-insulated, stainless steel sheathed thermocouples and RTDs can be bent to fit a specific installation. This is also true of the hard line cable used as a transducer extension cable for high-pressure, high-temperature applications. The manufacturer will specify a minimum bend radius for the temperature probes or hard line cable. Tighter bends can weaken or fracture the steel sheath and possibly break the internal wires.

That threads and components be correctly lubricated is of special importance when the devices are used at higher temperatures for an extended period. Using the appropriate lubricant will enable the transducer to be easily removed after its exposure to high temperature.

Lack of lubrication can cause damage to a transducer when excessive torque or force is needed when installing or removing it.

The transducer user is encouraged to carry out any possible tests to verify that the transducer has been installed successfully. This can include performing insulation testing of the strain gauges prior to wiring, applying a transducer stimulus to verify the wiring and operation of signal conditioners, and even in situ calibration of the complete system.

7
Data Acquisition

The data acquisition or display section of a measurement system includes the equipment that connects to the output of the signal conditioning equipment and processes that signal into a form that the system operator can use. The equipment can be as simple as a digital meter or as complex as a computer-based data acquisition system. Which type is chosen depends on such factors as the number of signals to be accessed, how fast the data is changing, whether the data needs to be stored, and whether the data needs to be processed further.

A measurement system may contain several different types of displays depending on the operator's needs. For a given test, the operator may need to monitor the pressure, temperature, and flow of a fluid in a test rig. At the same time, the operator may need to store vibration measurements and perform online analysis of the vibration data. The monitoring could be done using digital indicators or a data acquisition system, while the vibration measurements could be carried out using a separate vibration analysis data acquisition system.

7.1 Digital Indicators

Digital voltmeters are the simplest type of digital indicator. These units measure the signal conditioner's DC voltage output signal and display the result in DC volts. A pressure transducer connected to a signal conditioner may provide a linear output of 0 to 10 volts DC when the input is changed from 0 to 100 psig. The signal conditioner output is connected to a digital voltmeter that has a range of 0 to 10 volts DC. This display will show a reading of 0.00 when the pressure is 0 psig and a reading of 10.00 when the pressure is 100 psig. The operator knows that the reading is in volts and not in psig and must mentally convert the voltage reading to a pressure. In this case, the conversion is 10 psig per volt, which makes the mental conversion process easy. A conversion of 32.5 psig per volt would make the operator's job somewhat harder. If the output-to-input relationship is not linear then he or she would have to depend on a lookup table to determine the pressure.

Digital indicators that have a scalable display are the next most complex. These units have either manual or programmable adjustment capabilities so the display can be made equivalent to the engineering units' value measured by the transducer. This is usually done with a combination of decimal point location and a gain or attenuation scaling adjustment. For the linear 0 to 100 psig transducer all that is required is a shifting of the decimal point one place to the right. The display will now indicate 00.0 for an input of 0 psig and 100.0 for an input of 100 psig. The 32.5 psig per volt input will likely require a decimal point shift and a scaling adjustment to make the display indicate directly in engineering units.

Digital indicators are also available that have built-in linearizing circuits. These can be fixed circuits designed to linearize a standard transducer output such as a thermocouple or RTD. They can also have adjustable circuits that can be set up to match a custom nonlinear transducer. This can be achieved by approximating the output curve with a series of straight-line segments. The user has control over the segments' slopes and end points (break points). The linearization can also be performed by a microcomputer built into the display. In this case, the linearization is accomplished with a program or lookup table. Figure 7-1 illustrates a digital voltmeter with a resolution of eight-and-a-half-digits. Figure 7-2 illustrates a digital voltmeter with a resolution of six-and-a-half-digits.

Other features that can be built into digital indicators include dual displays, multiple switch selectable inputs, data storage over a selectable

Figure 7-1. Eight and One-half Digit Multimeter
(Courtesy, © Agilent Technologies, Inc. 2000
Reproduced with Permission, Courtesy of Agilent Technologies, Inc.)

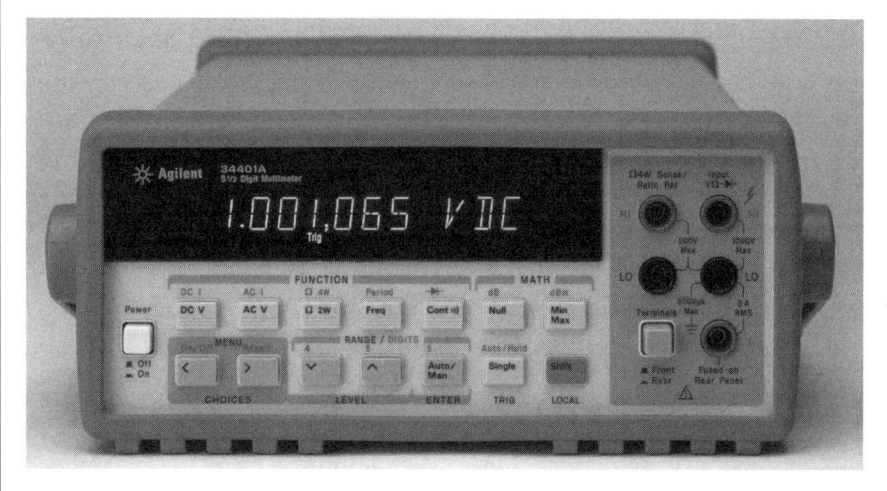

Figure 7-2. Six and One-half Digit Multimeter
(Courtesy, © Agilent Technologies, Inc. 2000
Reproduced with Permission, Courtesy of Agilent Technologies, Inc.)

time period, retransmission of an analog signal that is proportional to the digital reading, filtering, and alarm or trip outputs.

The resolution of a digital display is dependent on the number of digits used in the display. A three-digit display can indicate any value from 000 to 999, or one thousand possible values. The resolution is thus one in a thousand, which represents 10 millivolts for a 10 volt input signal or 0.1 psig for a transducer range of 0 to 100 psig. The three-and-a-half-digit display has a fourth leading digit that can be either 0 or 1, and this display can indicate any value between 0000 and 1999, or two thousand possible values. The resolution is now one in two thousand. However, for a direct reading of a 10 volt or 100 psig signal we haven't increased the display resolution since one half of the possible values will not be used. To increase the resolution in this case we would have to use a four-digit or four-and-a-half-digit display. Increased resolution is meaningless if the value of the least significant digit is less than the error value of the input signal or the error of the display. Resolving a pressure signal to 0.01 psig is meaningless if the accuracy of the signal is only ± 0.1 psig. Increased resolution also means that the least significant digit may fall within the noise value of the signal. In this case, the last digit of the display will fluctuate so quickly and randomly as to become useless.

Digital displays are packaged in a number of different formats including bench mounts, panel mounts, and handheld devices.

7.2 Recorders

Recorders are data acquisition devices that preserve the measurement signals on paper, usually in a graphical format. They can have a single input or up to ten or twelve inputs and can have fixed or variable paper speeds. In typical recorders the paper moves on rollers at a constant speed, and the pens move across the width of the paper at a distance that is proportional to the amplitude of the input signal. The recording is thus a visual presentation of how the signal varies with time (X-T recorder). Figure 7-3 shows a typical X-T recorder. Other recorders present the signals from two sources on a sheet of graph paper with one signal on each axis. These are called "X-Y recorders" and visually show how one signal reacts with respect to another during a common time period.

Recorders are the precursors to data loggers and data acquisition systems and have largely been replaced by these devices. Some hybrid recorders can combine the data storage features of data acquisition systems and the graphing capability of recorders. These units can store a large quantity of high-frequency data in a digital format and then reproduce it using the pen and chart features of the recorder.

7.3 Data Loggers

Data loggers are standalone units designed to acquire and store a large amount of data, usually from a large number of transducers. They are

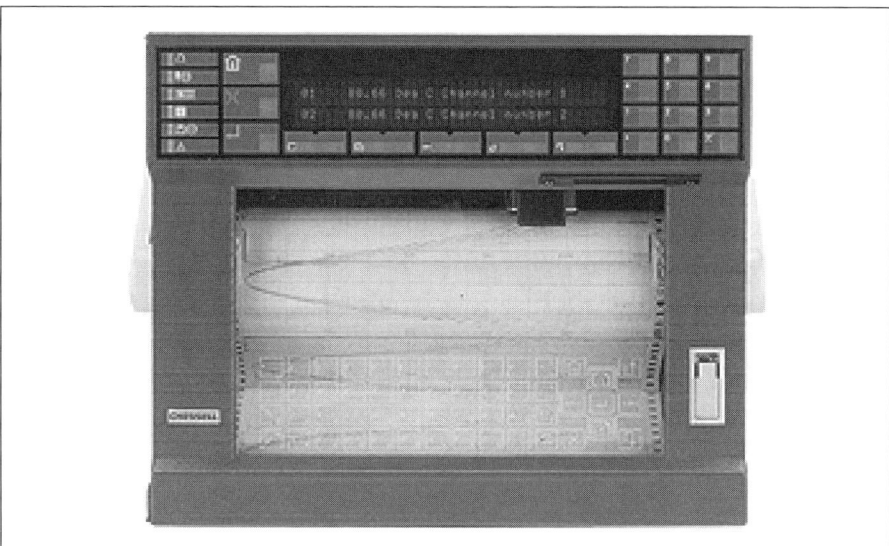
Figure 7-3. Multi-channel X-T Recorder
(Courtesy, Eurotherm Chessell Recorders)

then able to present the data in a number of preprogrammed formats such as strip charts or tabular columns of data that are identified by the time and date of acquisition. These units were designed to archive test or process data over a long period of time. They do not have the same amount of programming flexibility as a data acquisition system and are usually designed to have a limited number of acquisition modes and data presentation modes.

A number of modular data logger "boxes" are also available that access a limited number of sensors, store data, and then dump the data to a personal computer via the serial or parallel ports. These units typically have the transducer signal conditioning built in and can be battery powered for use in remote areas. A software package is supplied with the logger to allow the computer to download the data and then carry out a choice of preprogrammed data presentation formats.

7.4 Data Acquisition Systems

Computer-controlled data acquisition systems have become the most popular form of equipment for interfacing with transducers or signal conditioners and then storing, manipulating, and displaying the test data. These types of systems proliferated around 1980 when the personal computer became readily available at a reasonable price. Before this, the data acquisition systems were controlled by expensive mini computers. Systems with extensive capabilities, high speeds, and large storage capabilities were developed as computer processing power and speeds increased and memory costs decreased.

7.4.1 Personal Computer–Based Systems

The personal computer or PC-based data acquisition system is usually designed and assembled by the user out of a wide variety of hardware components and software that integrate with the computer to form the system. The hardware can either be installed within the computer enclosure or in external boxes. The internal system uses printed circuit cards that plug directly into the computer bus slots. The external system houses the cards and transfers data and control signals via the computer's external serial or parallel ports.

The system's overall capability to transfer large amounts of data and/or perform real-time analysis of high-frequency signals depends on the computer's capabilities, the data acquisition hardware, and the control and application software. Almost any personal computer is capable of taking a few measurements per second from a small number of transducers and displaying the value of the signals on a monitor. The computer requirements for continuous high data throughput and real-time analysis include a 32-bit processor with coprocessor, a high-speed bus, and direct memory access (DMA) capability. A second high-speed hard drive for data storage may also be required, while the first drive contains the operating system and data acquisition software. The USB port and PCMCIA bus do not handle direct memory access, and the RS-232 and RS-485 serial communications are limited in transfer speeds. PCI or IEEE-1394 buses are the best choices for high storage rates and real-time processing.

The data acquisition system hardware includes as a minimum a multiplexer and an analog-to-digital converter. Other functions that may be carried out with the hardware include amplification, isolation, linearization, filtering, and supply of transducer excitation voltage. These additional functions may be required depending on the level of signal conditioning that the rest of the measurement system provides. The hardware must be capable of DMA transfers if high data throughput rates are required.

The multiplexer is a series of switches that route many analog inputs one at a time to a single analog-to-digital converter. The switches are typically solid-state (transistor) devices, although electromechanical switches are used for special applications. The multiplexer inputs can be single-ended or differential, with the same constraints as in single-ended or differential input signal conditioning. The multiplexer's switching speed and the conversion time of the analog-to-digital converter will determine the sampling rate of the data acquisition system. The number of readings of a given channel will be inversely proportional to the number of channels attached to the one A/D converter. A 100-channel system with a sampling rate of 100,000 samples per second will only read a given channel once every millisecond. The time skew caused by taking readings from a large number of channels at different moments can be eliminated by using a sample-and-hold circuit on each channel. This circuit uses switches to take a reading from each input at the same time and store the reading in a capacitor. The multiplexer then sequentially scans the stored readings and transfers them to the A/D converter.

The important characteristics of the A/D converter include conversion time and resolution, which must be selected based on the test requirements. Other system hardware includes digital-to-analog converters, digital inputs and digital outputs, external and internal triggers, and timing circuits.

The PC system's software requirements include the driver software that controls the data acquisition system hardware, the computer's operating system, the high-level languages supported by the driver software, and the application software that is available. These must be compatible with each other and provide the necessary features to carry out the measurement requirements.

Plug-in Cards

Data acquisition systems can be assembled using printed circuit boards that reside within the computer enclosure and plug into the computer bus. This makes possible a very compact system with a minimum of external wiring and cabling. This type of system is limited by the number of boards that can be plugged into the computer, and it exposes the analog signals to the very noisy digital electronic environment within the computer enclosure.

External Box System

Larger data acquisition systems may require that a separate enclosure be used to house the number of boards required. This is especially true if more of the signal conditioning functions are part of the data acquisition hardware. Separate boards may be required for signal amplification, filtering, and sample-and-hold circuits. This enclosure is connected to the computer through serial or parallel ports or through a bus extender card and cable. Physical separation from the computer enclosure can create a less noisy environment for the analog signals. This is valuable when low-level transducers such as thermocouples are connected to signal conditioning within the data acquisition hardware.

7.4.2 Manufactured Systems

Manufactured systems are designed from the ground up to be dedicated data acquisition systems rather than add-ons to commercial personal computers. This allows the manufacturer to have complete and optimal control over the physical layout of various functions, to have separate computing power where needed, and to develop powerful application software. This software can either cover a wide range of data acquisition requirements or be tailored specifically to one major type of measurement such as strain gauges or vibration of rotating machinery. The control computers for these systems will have features not found in desktop personal computers. These features can include extra cooling, improved air filtering, shock mounting, and redundant components such as power supplies, disk drives, and even processors. These features can be very important if the system is to be used in a lab or manufacturing environment that has less than ideal atmospheric and vibration conditions.

These types of systems are more expensive than plug-in systems, but the user may save in the long run by having an integrated, tested product that is designed by a single knowledgeable system designer. This contrasts with the plug-in system where the user is responsible for integrating hardware and software that may not be from the same manufacturer and may not be totally compatible.

7.5 Filters

Filters are an important part of the data acquisition system hardware and are needed to remove noise and unwanted frequencies from the measurement signal. The filters can be any of the types described in section 5.5, "Filters," but the usual type is the low-pass filter that has a corner frequency set to remove high-frequency noise or unwanted high-frequency signals. The filters can be installed on a channel-by-channel basis, or one single filter can be installed before the A/D converter. The channel-by-channel system allows the user to set corner frequencies and filter types depending on the individual channel requirements. In many cases, the filters can be set up using the application software provided with the data acquisition system.

When measuring AC or periodic signals such as vibration signals, the data acquisition system must contain an anti-aliasing filter. This is a low-pass filter with a very sharp amplitude roll-off at frequencies above the corner frequency. The sampling or Nyquist theorem states that the sampling frequency must be at least twice the highest frequency in the data being sampled in order to accurately reproduce the digitized signal. Another way of stating this is that the highest frequency in the measured signal can be no more than one-half the sampling frequency. If this condition is not met then a low-frequency alias signal will appear in the reconstructed signal. This alias signal will distort the result since it includes a component that was not present in the original signal.

The anti-aliasing filter prevents signals with frequencies greater than one-half the sampling frequency from entering the system. Either one filter per channel or a single filter prior to the A/D converter is required depending on whether a variable sampling rate per channel or a constant sampling rate for all channels is used. The system user must know the frequency content of the signals to be measured or decide on a frequency limit in order to set an appropriate sampling rate and filter corner frequency before taking measurements.

8
Data Reduction and Analysis

Data reduction and analysis is the process by which large amounts of raw or partially processed data obtained during a test is managed and organized and the results presented in a visual format. The purpose of the data reduction and analysis process is to pass on the information obtained during the test so that conclusions can be made about what happened, about signal validity, and about the need for more or different tests. Data reduction and analysis can take place in real time, after the test, or a combination of both. Almost all data reduction and analysis is carried out using computer programs to store, manipulate, and present data.

The simplest form of data reduction is the conversion of signal conditioning voltage outputs into engineering units. This can involve feeding the voltage into an equation that describes the output-versus-input relationship as determined from the calibration of the measurement system. In other cases, the outputs from several measurements are combined to produce one result that is then used in later analysis. The use of temperature, pressure, and volumetric flow signals to produce a mass-flow measurement is an example of this type of data reduction.

Another simple form of data reduction is correcting measurement signals for the errors caused by environmental or test conditions that affect the transducer signal. These can include zero shifts, span errors, and dimensional variations caused by changes in temperature or pressure. The correction factors for these effects must be known before the data reduction procedure, either from manufacturer's specification sheets or from bench tests on the transducer. The necessity of measuring parameters that affect transducers must be anticipated when the test plan is designed so this type of data reduction can be conducted.

The corrected engineering unit data is stored in new files and used in further data reduction steps such as graphing and sorting, as described in the next section. The original raw data should always be preserved so the user can revisit it for any reason.

8.1 Graphing

Graphing is the primary method for presenting test data in a visual format. The viewer of the graph can quickly scan the results and see trends, peaks, valleys, anomalies, and correlations. If a picture is worth a thousand words then a graph is certainly worth a thousand data points.

The line graph can be used to show how the measurements changed over time. It is simply an X-Y graph with the amplitude of the signal plotted on the y-axis and the acquisition time plotted on the x-axis. The individual points are joined by lines to indicate how the signal varied continuously over a certain time period. Joining the data points with lines is a way of estimating what happened between recorded data, with no guarantee that an interpolated value would actually fall on the line. The more points recorded in a given time, the greater the likelihood that the graph truly represents the continuously varying signal.

One or more sets of data from other signals recorded at the same time can be added to the same graph. This may visually illustrate some time relationships between the signals, either by showing common increases or decreases in signal amplitudes or by showing signals changing at a common rate.

The line graph with time as the x-axis is useful for depicting vibration data since it directly shows us the most important vibration parameters. If the signal is from a displacement transducer then the y-axis shows us the vibration amplitude. The frequency can be determined by the time between displacement cycles, and the velocity can be estimated from the slope of the line between points.

The scatter graph, which is similar to the line graph, can be used to show the relationship between a series of measurements that were recorded at the same time. One example is a test conducted to determine the deflection of a component as the flow past it is increased. The graph would have the flow plotted on the x-axis and the deflection plotted on the y-axis. In this case, a straight line or a curve fitted through the points would show the relationship.

The histogram is a graph that shows the distribution of a large number of repeated results of the same measurement. The x-axis shows the total range of results from the minimum to the maximum divided into several equal-width bins. The x-axis shows how many results fell into each bin. The expected result is a graph in which the tops of the bins show a bell-shaped or normal distribution. Anomalies such as data skewed to one side, double peaks, or multiple peaks could indicate problems with the data.

Other types of graphs include bar and column graphs, 3-D surface graphs, contour graphs, and area graphs.

8.2 Sorting

Sorting is a process that examines a large amount of test data and extracts a few values that may give us significant information about the test results. These values may be all that we need to make conclusions about a test.

8.2.1 Minimums
Finding the conditions that cause a measurement to achieve its smallest value is often the goal of a product development test. Manufacturers will often want to know the optimum design for the least vibration or noise level in their product. A scan of the data for the minimum value will find the contributing conditions.

8.2.2 Maximums
The maximum data values are often associated with conditions that will cause damage to components or will shorten a product's life. Locating these maximum values and identifying the contributing factors from other data measured concurrently enables designers to modify conditions and reduce values.

8.2.3 Medians

The median value in a set of data points is the one that has the same number of points greater than and less than the value. If a set of 75 data points is put in ascending order, then the 38th point is the median value, and there are 37 values smaller than the median and 37 values greater than the median. The median value may be closer to the typical value than the mean value for the case where the data distribution is skewed instead of normal.

8.2.4 Means

The mean value in a set of data points is the average of all the values in the set. The mean or average is found by summing the values of all the points and dividing the sum by the number of points. For a normal distribution, the mean is the value that is most likely to occur. Averaging sampled data will reduce random noise since the random values will not occur at each sampling, while the constant value will appear in each sample.

8.3 Signal Analyzers

Signal analyzers are data acquisition systems that include additional software designed to process and manipulate data in a variety of different formats. These formats display the results in a way that is specific to the type of analysis being carried out. Analysis types include strain gauge, audio, vibration, and rotating machinery. These analyzers can be portable units with two- three- or four-channel capability or multi-channel rack-mounted systems that are capable of interfacing to hundreds of measurement sensors.

Strain gauge or stress analysis systems are primarily designed to process measurements from strain gauges and produce stress information about the component or material being tested. The systems also accept information from other types of sensors such as thermocouples or pressure sensors that may be used in conjunction with the strain gauges. The software accepts information about each channel's specific strain gauge, including gauge type, resistance, gauge factor, temperature coefficient, and excitation voltage. The software uses this information to complete the required data reduction, including temperature compensation, bridge linearization, and conversion into engineering units. The software also mathematically processes the data to calculate stress values from individual gauges and to resolve signals from two- and three-gauge rosettes. Systems are available for dynamic strain measurements as well as static measurements. A 40-channel dynamic system capable of 200,000 samples per second would allow 5,000 readings per second per channel.

Audio analyzers are designed to handle sound or noise measurements in the audio frequency range from 20 hertz to 20,000 hertz. These systems measure sound pressures, sound intensities, and sound distributions. They are used to develop and test audio products and to monitor equipment sound levels and the effects on the human ear.

Vibration analyzers monitor the periodic motion of components using the outputs from sensors such as displacement transducers, velocity transducers, and accelerometers. The software determines vibration amplitudes and the frequency distribution of the signal as well as identifying multiples and submultiples of the fundamental frequency. Vibration analyzers perform mathematical functions such as integration and differentiation to change signal units from acceleration to velocity or displacement and vice versa. Frequency response tests can be carried out to determine how the amplitude and phase shift of a component measurement will change with respect to the frequency of a driving signal. These types of tests are conducted on components, structures, and systems to find the causes of destructive vibrations or to predict how certain vibration frequencies will affect the test object.

Rotating machinery analyzers use software to monitor the vibration signals from equipment such as motors, pumps, and generators. Much of the measurement data is analyzed with respect to each revolution of the rotating shaft. The software is designed to identify conditions, such as worn bearings and shaft misalignments, that can eventually lead to equipment breakdown.

8.3.1 Time Domain Analysis

Time domain analysis is performed by observing the amplitude and phase of the signals with respect to time. The information displayed is like that shown on a strip chart recorder or an oscilloscope. This type of display has an intuitive feel to it since it visually shows what is taking place. In vibration studies, the time domain is useful for showing impulse-type signals caused by bearing or gear problems and for showing apparently clipped signals caused by looseness in components. Once the signals become complex and nonsinusoidal, the amount of quantitative information that can be obtained becomes small, and it is better to use the frequency domain for analysis.

8.3.2 Frequency Domain Analysis (Fourier Analysis)

Frequency domain analysis is conducted by observing the amplitude and phase of the signals with respect to frequency. This technique is based on the fact that any complex periodic signal can be broken down into a number of sine waves of varying amplitude and frequency. A square wave can be broken down into a sine wave at the same frequency and amplitude as the square wave, plus a sine wave at 3 times the frequency and 1/3 the amplitude, plus a sine wave at 5 times the frequency and 1/5 the amplitude, plus a similar continuation of all the other odd harmonic frequencies. The mathematical algorithm that produces this conversion from the time domain to the frequency domain is called the Discrete Fourier Transform, and the approximation of this algorithm that is used in signal analyzers is called the Fast Fourier Transform or FFT. Being able to see all the component frequencies and their amplitudes provides much more information to the user than the time domain display provides.

8.3.3 Real-Time Analysis versus Post-Test Analysis

Real-time analysis occurs when test data is acquired and stored as time domain files in the background while data analysis and the display of results is occurring in the foreground. This allows the operator to monitor the test conditions and obtain analyzed results as the test progresses. Under these conditions, the operator may notice anomalies or unexpected results that can be investigated further in post-test analysis or in additional tests.

Post-test analysis is conducted on test data that was stored by a data acquisition system while the test was being carried out. This would occur when the test duration is too short for online analysis or when the online analysis would prolong the test for an unacceptable length of time. This method also allows the operator to run different types of analysis on the same data.

9
Equipment Calibration

Calibrating all the equipment used in the test measurement system is an essential step in the larger process of obtaining valid test data. Calibration involves applying a series of known inputs to the equipment and recording the resulting output values. The calibration process may also involve adjusting the equipment to eliminate or minimize the errors that can occur in these output values.

To understand the importance of calibration the following points should be kept in mind:

- Calibration of all the equipment used in the test measurement system must be performed. This means that every piece of instrumentation that is installed to set test conditions or to make resulting test measurements must be calibrated. Any measurement that is cited in the test results or used to make test conclusions must have been made with calibrated instrumentation.
- Calibration is an essential step because it is the only thing that can guarantee the validity of the measurements and the test conclusions. Calibration allows us to assign valid numbers to the measurements and to state the possible error in those measurements. Without calibration, all that could be reported are relative results such as "there was an increase in vibration as the flow increased."

- When we say that calibration involves applying a "series of known inputs" we mean that the instruments are calibrated over the full range of measurements that are taken. Any measurements that are taken outside the calibrated range cannot be considered valid since there is no guarantee that the instrument would have produced those specific results. A series of inputs is also required to increase our confidence in the interpolated values that fall between each of the known inputs. Calibration must be performed at a sufficient number of steps in order to determine instrument linearity or conformance.
- Using known inputs means that the value of the source or signal used in the calibration is known to be true within some specified tolerance. We use calibration standards to serve as a reference when producing these known inputs. These standards have a substantially smaller error than the test instruments and have themselves been calibrated by more accurate standards.
- The recording of the outputs during calibration means that documentation is used that specifies the values of the output for the series of known inputs. This documented information allows us to determine the degree of error in our measurements and serves as proof that the calibration took place.
- The fact that the calibration is described as a "process" means that it is a planned activity that is carried out using a prescribed procedure. The procedure specifies how, when, and under what conditions the calibration took place. The calibration process is often a requirement of a manufacturer's quality assurance program, which will document the activities to be carried out.
- We may adjust the output of the instrument being calibrated if such adjustments are available. A device such as an RTD does not have adjustments, and the calibration involves recording the output resistance for a series of known input temperatures. A pressure transducer with integral amplifier will likely have zero and span adjustments available. These adjustments can be used to set the output as close as possible to a nominal output

such as 0 to 10 volts DC when a series of known pressures are applied to the input of the pressure transducer.

Each test laboratory and manufacturer will have to decide whether to carry out calibrations in house or have them done by an outside calibration service contractor. This decision is based on the cost of each choice and an evaluation of any factors that do not have a direct cost but do impact the operation of the facility. The final decision may be a combination of the two choices, where some equipment is calibrated in house and some is contracted out.

Some of the questions to consider when making this decision include:

How many instruments are to be calibrated?
A small number of instruments will not likely justify the cost of in-house calibration. A large number of instruments will allow the organization to spread the equipment, facilities, and manpower costs, resulting in a lower per-calibration cost. At some volume, the in-house calibration cost should be less than the outside contractor cost.

Are calibrations carried out on an "as required" basis, or are the instruments required to be in constant calibration?
A manufacturing facility will likely have a fixed inventory of instruments permanently installed on the plant equipment. The manufacturing process is continuous, and all the instruments are required to be in a calibrated state at all times. A test laboratory may have a large inventory of instrumentation to be used on a variety of tests and a number of different test rigs. At a given time, a large percentage of the instruments may be in storage or installed on inactive test rigs. In this case, it may be more cost effective to calibrate on an "as required" basis so as to reduce the volume of calibrations. The "as required" scenario with its smaller volume may still justify the in-house calibration since the required use of instruments is less certain and the outside contractor may not be able to carry out calibrations in the time required.

How often are the instruments calibrated?
The calibration period is determined by the type of instrument, the stability history of the individual instrument, the extremes of environment and use, and the manufacturer's recommendation. Deciding on

the calibration period is a form of risk management that requires careful assessment. A manufacturer must weigh the cost of more frequent calibration against the potential cost for product recall if instruments are found to be out of tolerance when recalibrated. The test laboratory must consider the cost of repeating tests or the effects of issuing incorrect test data. In some cases, a laboratory will calibrate instrumentation before and immediately after a test or series of tests. This is the only way to guarantee that the calibration of an instrument has not changed during the period of use. The before and after calibration is mandatory in situations such as the qualification testing of equipment to be used in nuclear power stations. When carrying out pressure tests on equipment it is necessary to calibrate the instrumentation within a specified time prior to the test or series of tests.

How many types of instruments are used in the facility?
A facility may require that only one parameter, such as pressure, needs calibrated instrumentation. Another facility may need to calibrate pressure, temperature, flow, and vibration. Each type of measurement may require unique standards, facilities, and staff capabilities. Providing multiple in-house capabilities increases some costs, but the use of common requirements such as software, auxiliary equipment, space, and manpower may provide savings.

What ranges of measurements are required?
Most standards are designed to cover only a certain measurement range, but a facility may be required to make measurements over a very wide range. Pressure instrumentation used in one facility commonly covers ranges from a few inches of water to many thousands of psi. This situation will result in several pressure standards being used as well as the higher costs associated with purchasing and maintaining multiple standards. Again, some savings may occur as a result of common requirements being used.

Is staff available to carry out calibrations in-house?
The availability of personnel is a big factor in deciding whether to perform calibrations in-house. Things to consider include:

- Experience and skills
- Required training
- Full-time job or additional duty

Are facilities available?
The facility requirements will likely be dictated by the standards used for calibration. If the standards can be used at the instrument location then a small storage and maintenance area is the only requirement. To gain higher-accuracy calibrations, a controlled environment calibration laboratory with extensive auxiliary equipment may be required.

What time period is involved?
If calibration is an ongoing, long-term requirement then the costs associated with purchasing standards, providing facilities, and training staff can be spread over several years. Most standards have a relatively long life and would only have to be replaced every ten to twenty years.

What is the cost of maintaining standards?
All standards require some level of routine maintenance, ranging from simple cleaning to replacing worn parts and operating fluids. The major expense is for periodic recalibration of the standards, which is usually carried out by a calibration standards laboratory. Recalibration must usually be done annually, at costs ranging from several hundred to several thousand dollars per standard.

What is the cost of training?
Staff will have to be trained in how to operate equipment, use calibration techniques, and operate and maintain the calibration system software and database. Some training can be provided by more experienced staff while other training will have to be obtained from equipment suppliers.

What is the cost of maintaining the calibration system software and database?
This work includes preparing calibration procedures, calibration reports, and instrument recall notices. It also includes maintaining and backing up files and installing software updates.

What is the cost of the required standards?
Standards tend to be high-cost items due to the tight manufacturing tolerances, the amount of labor-intensive assembly and testing, the cost of materials, the high development costs, and the low production runs. The investment must be able to be spread over a number of years, and there has to be a relatively high volume of use to justify the expense.

What is the calibration turnaround time?
The total time required for calibration has to be considered if the instrumentation is to be sent to an outside agency for calibration. This includes the time required for removal, packing, shipping, receiving and reinstallation as well as the actual time to carry out the calibration. The cost of spare replacement instrumentation has to be considered if the turnaround time is greater than any allowed downtime in the work process. These considerations also apply to having in-house standards calibrated by outside laboratories.

What are the packing and shipping costs for outside calibration?
The packaging and shipping costs depend on where the calibration agency is located and the amount of protection the instrument needs during shipment. A local agency means low shipping costs and a minimum of protective packaging if the instrument can be hand delivered. A shipment across the country will be more expensive, and special packaging may be required to eliminate shock, vibration, and temperature and pressure extremes that could damage the instrument or affect the calibration.

What is the cost of outside calibration service?
The cost to obtain outside calibration services must be known if a comparison with in-house calibration is being made. The outside contractor may have the advantages of large volume, lower labor rates, and the fact that calibration is his prime business, but he still has to make a profit. The number of non-cost considerations may tip the balance in favor of in-house calibration.

Other considerations include:

- Availability of qualified calibration contractors
- The risk of outsiders knowing your business
- Can you sell your calibration services?
- Some in-house expertise in calibration will still be required
- Calibration costs should be considered as part of the product development costs. The effect on product price must be evaluated.
- Degree of confidence with results from outside – shipping and handling effects
- Removal and reinstallation costs
- The shipping and handling effects if you purchase calibrated instruments

9.1 Component Calibration

Component calibration involves separately calibrating each component in the measurement system, usually carried in a calibration laboratory. The separate components could consist of the transducer, the signal conditioner, and the display device.

The components that make up a temperature measurement system might consist of an RTD, a linearizing amplifier, and a digital meter. In this case, the RTD would be calibrated in a temperature bath with a standard RTD as a reference and a resistance bridge or high-accuracy resistance meter used to measure the RTD resistance. The linearizing amplifier would be calibrated with a decade resistance box or RTD simulator connected to the input and a high-accuracy digital voltmeter connected to the output. The digital display would be calibrated using a variable voltage source and high-accuracy digital voltmeter connected to the input, and the resulting digital display would be recorded as the output reading. During these calibrations, the output of any component with zero and span adjustments would be adjusted to be as close as possible to the ideal nominal output. The overall system accuracy would have to be calculated from the errors found in each separate component calibration. Once calibrated, the components would be installed and wired in the test rig or plant.

The advantages of component calibration include:

- Information on the performance of each separate component is available
- Offers greater opportunities to optimize the overall system accuracy
- Individual components can be replaced with other calibrated components if necessary

The disadvantages of component calibration include:

- The calibration process is more time consuming
- Installation and wiring effects are not seen or accounted for
- The system error is calculated rather than measured
- More standards are required

Component calibration may be performed the first time a measurement system is assembled, while subsequent calibrations may be system or in situ calibrations as described in the following sections.

9.2 System Calibration

In a system calibration the individual components of a measurement system are wired together as they would be in the final field installation. The assembled system is then calibrated, usually in a calibration laboratory. For the temperature measurement system described in section 9.1, the RTD would be inserted into a temperature bath along with a standard RTD. The calibration would consist of varying the bath temperature and noting the standard RTD reading and the digital display reading.

The advantages of a system calibration include:

- The calibration process is less time consuming
- Fewer standards are required
- The system error is known rather than calculated

The disadvantages of a system calibration include:

- The calibration procedure is more complicated
- The individual components must be recalibrated if the system is dismantled
- A new system calibration must be done if any one component has to be replaced

9.3 In Situ Calibration

To perform an in situ calibration, all the components of the measurement system are installed on the test rig or plant equipment and wired as they would be in the final configuration. A system-type calibration is then completed in the field by applying known inputs to the transducer and recording the resulting readings on the display.

The advantages of an in situ calibration include:

- All installation and wiring effects are included in the calibration
- Any environmental effects on the components will be included in the calibration

The disadvantages of an in situ calibration include:

- Two people may be needed to perform the calibration if the distance between the transducer and display is too large
- Communications equipment may be required if two people are performing the calibration
- Requires standards that can operate in a field environment
- May require platforms and ladders so personnel can reach difficult transducer locations

9.4 Calibration Standards

Calibration standards are the reference instruments that provide or measure the known inputs to a transducer during the calibration process. They have a higher accuracy than the transducers and have themselves been calibrated by higher-accuracy standards on a periodic basis. The accuracy of a standard should be between three and ten times the accuracy of the instrument being calibrated in order to minimize the effect of the standard's error on the overall calibration error.

Standards are designed to have a high degree of stability and repeatability in order to provide long-term accuracy. A standard may be in daily use and perform hundreds of calibrations before being returned to a standards laboratory or manufacturer for recalibration after a year or more. The standard's error should still be within the manufacturer's specification at the end of the calibration period as long as it has been handled and maintained as required.

Standards must be handled or used with care since they are usually more fragile and sensitive to the environment than the measurement instrument they calibrate. For example, the zero resistance of a Standard Platinum Resistance Thermometer (SPRT) can be shifted just by tapping the SPRT on a hard surface. The mass of precision dead-weight tester weights can be altered by rough handling or even exposure to the fatty acids in our skin. The standards should be kept and used in the environment for which they were designed. Standards that were designed for use in a calibration laboratory should not be used for field calibrations because environmental extremes may affect the standard's accuracy or alter its calibration.

The accuracy of the standards should match the accuracy of the measurement instrumentation being calibrated. Using standards that are more than ten times more accurate than the instruments being calibrated will result in higher costs due to the purchase price of the standard, the extra time required to complete high-accuracy calibrations, the higher maintenance and recalibration costs, and the additional operator training that is required. The overall measurement uncertainty will also not increase since other transducer characteristics will override the increased calibration accuracy.

There is a hierarchy in the relative accuracies of standards. These range from the national or international standards maintained by a country's national calibration laboratories to the working standards used by facilities to directly calibrate the measurement instrumentation. These standards are described further in section 9.11.

9.5 Multifunction Calibrators

A multifunction calibrator is a working standard that is capable of calibrating more than one measurement parameter. These can be calibration laboratory calibrators or field environment calibrators. They can often cover more than one range by using multiple internal standards or multiple plug-in or external modules.

An example of a laboratory type multi-calibrator would be one that is designed to calibrate electrical parameters such as AC and DC voltage, AC and DC current, resistance, capacitance, and frequency. These units can be accurate enough to calibrate six and one-half digit multimeters throughout their complete ranges and functions.

An example of a field type multi-function calibrator would be one designed to calibrate pressure transducers and transmitters. These units have one or more pressure standards and high-accuracy voltage and current measuring capability and are able to provide excitation voltage to the transducer or transmitter. They are usually keypad programmable and are able to display the applied pressures in a variety of units as well as store calibration results for later downloading into the calibration system software.

9.6 Calibration Records

Calibration records document all relevant details concerning each calibration that is carried out in a facility. The records are controlled and stored according to the facility's quality assurance program and are made available to various groups as needed. The instrument user needs to know that the calibration was successful and will use the calibration results in the data reduction phase of the work. The quality assurance personnel will require the records during internal and external audits as proof that the calibrations were carried out. The calibration staff uses the records to schedule the next calibration and to determine the long-term stability of the individual instrument so that an appropriate calibration period can be assigned.

The records answer the questions how, when, where, and who and detail the conditions during calibration and the calibration results.

The calibration procedure provides the details as to "how" the calibration is to be carried out for a specific instrument. The quality assurance manual will likely describe the general steps that are to be followed for every calibration. The procedure will specify the standards that are to be used, describe the equipment setup, and advise of any preparation or warm-up requirements. It then describes the step-by-step calibration process and may also detail the method of adjustment if required. Having a written procedure helps ensure that calibrations are carried out on a consistent basis independent of different equipment operators.

Information about "when" should include the time and date of the calibration, the time and date when the instrument is due for recalibration, and the time and date of the standard's calibration and recalibration.

Details about "where" should include the calibration location and the location where the instrument is to be used or stored after calibration.

The latter information is important when the time for recalibration comes due and the instrument must be located.

Information about "who" includes the person who carried out the calibration, the person who authorized the calibration results, and the person who is using and controlling the instrument after calibration. Knowing the user of the instrument also makes it easier to locate the instrument when recalibration is due.

The "conditions" under which the calibration was performed may include the room temperature, the relative humidity, and the atmospheric pressure. One or more of these conditions may affect the accuracy of the standards or the instrument being calibrated. If the conditions of use are different from the conditions of calibration the user may have to take this into account during the data reduction phase. The calibration room conditions should be constantly recorded to demonstrate their stability. Some standards are sensitive to the rate of change of the room conditions as well as to the absolute value of the conditions.

The "calibration results" include the output from the instrument in response to the known input and the difference between the two, which is the error. This is usually shown in a tabulated format that includes all of the values that were obtained during the calibration. The error can also be calculated as a percentage of full scale or a percentage of reading and shown as part of the table. The calibration errors are compared to previously determined error limits that are based on the accuracy specification of the instrument. A calibration is considered a "pass" if all errors are within the limits and a "fail" if any errors are outside the limits. A failed calibration means that the instrument must be repaired or adjusted and then recalibrated. A subsequent failure means the instrument should not be used for the measurement and should likely be scrapped. The Test Uncertainty Ratio (TUR) or the Test Accuracy Ratio (TAR) can also be calculated for each calibration point to show that the standard has the necessary accuracy to be used to calibrate the particular instrument.

9.7 Calibration Laboratory Requirements

To properly perform instrument calibration a dedicated area with the appropriate environmental conditions, support services, auxiliary equipment, and working space is essential.

The degree of environmental control will likely depend on the standards used in the calibration process. The higher the accuracy of the standards the more likely that tighter control over the environment will be required. A national primary standards laboratory will likely require temperature control to be within 0.01°C for its coordinate measuring machine and laser interferometer room. Their other dimensional laboratories will be controlled to within 0.1°C. A laboratory with working class standards used to directly calibrate transducers will probably require temperature control in the range of 1°C to 2°C. Temperature is the most important condition that needs to be controlled, and steps should be taken to ensure that the level and stability are maintained. The laboratory should be located in the building's central area so there are no doors opening directly to the outside and no windows that allow excessive solar heating.

Humidity and vibration are other parameters that should be controlled. High room humidity will promote corrosion of metal surfaces and electrical contacts, while low humidity will increase the likelihood of damaging effects caused by static electricity. A relative humidity level of 40 percent with control to within ± 10 percent will likely keep the standards and their human operators quite happy. Measures should be taken to minimize any vibration caused by other equipment operating in the same building. This is most important for dimensional measurement laboratories, and it may require that isolated concrete pads be installed to support some measuring equipment.

Air filtration may be required to remove high levels of particulates from the laboratory atmosphere and to keep the equipment and work surfaces as clean as possible.

The design of a multifunctional calibration laboratory should incorporate separate rooms for different functions so that the activities in one area do not affect another area. A temperature calibration area will generate high ambient temperatures, fumes, and particulates that may require special air handling measures such as fume hoods, higher air circulation rates, and higher air cooling capacities. A vibration calibration area may include a shake table and provisions for isolating its vibration effects from other areas. Areas are also required for office work, paper file storage, auxiliary equipment storage, computer operations, and incoming and outgoing instrument storage.

The services that are required in a typical calibration laboratory include adequate single and multi-phase electrical supply, instrument air supply, and sinks with hot and cold water supply. Additional requirements include equipment for measuring and recording the temperature and humidity in the laboratory and in some cases the atmospheric pressure reading. Knowledge of the local acceleration due to gravity may also be required for pressure calibrations that use deadweight testers or for force calibrations.

9.8 Calibration System Software

The most productive way to control and utilize the data contained in the calibration records is to use calibration system software. The software that is available ranges from simple databases that store and retrieve calibration results to sophisticated systems that are flexible and powerful enough to carry out any requirement found in a modern calibration laboratory.

The first requirement of calibration system software is that it store information about the standards used in the laboratory and about the instruments that are calibrated by the laboratory. This information can include:

Asset Number

A numeric or alphanumeric identifier is assigned to each standard and instrument contained in the calibration system's database. The asset number is physically attached to the equipment by etching or by adding a printed label. The number is present on every document, report, and calibration record associated with the particular asset.

Type of Instrument

The type of instrument is a short combination of words that provides a functional description of the instrument. Examples could include "pressure transducer," "thermocouple," or "linear displacement transducer." This descriptive field allows the user to search the database for all instruments of this type and assign a generic calibration procedure to this group of instruments if desired.

Manufacturer

Having the name of the instrument manufacturer in the database allows the user to narrow searches to those particular instruments. This is useful for determining the calibration recall period for identical instruments or for deciding on the number of spare parts needed for maintenance.

Model Number

The model number is also used to narrow database searches to particular instruments.

Serial Number

The serial number is usually a specific identification of an individual instrument, but the number format varies from manufacturer to manufacturer and there is the possibility of duplication. An assigned asset number is the best way to identify a specific instrument.

Other information about each instrument includes the name of the individual or group using it and the current location. This information is used to find the instrument when it is due for recalibration. This type of information may have to change each time a particular instrument is calibrated.

Each calibration requires that the following information be stored: the calibration date and time, the recalibration due date, the calibration results, the ambient conditions, and the name of the person doing the calibration. The calibration results are used to calculate errors, determine TUR values, and determine whether the calibration was a pass or a fail. This information is combined with the previous information to produce documentation such as calibration reports and calibration certificates.

The calibration system software may also be able to generate custom calibration procedures or allow procedures developed by the software manufacturer to be used. These procedures can lead the operator through the calibration steps in real time and prompt him or her to enter calibration results using the computer. Operators can also perform computer-controlled procedures that control the standards and read or even adjust the outputs of the instruments being calibrated.

The data that is stored in the calibration database can be used to generate a wide variety of reports that are either designed by the software vendor or custom made by the software user. Powerful third-party software is often used to design these custom reports. Reports can include inventories, notification of instruments due for recalibration, and forward and reverse traceability reports. A forward traceability report identifies the chain of all standards used to complete the calibration of a particular asset. A reverse traceability report identifies all assets that have been calibrated by a particular standard since a given date or the date of its last calibration.

Some calibration software has the ability to interface with bar code readers, which are used to import data from bar codes attached to instruments, standards, and auxiliary equipment. This speeds up the data entry process and reduces errors caused by manual entry.

9.9 Software Validation

As more equipment and systems are controlled by computer software, it becomes increasingly important that the software be tested to assure that it operates as designed. This testing is known as "software validation" and can be considered as the equivalent of hardware calibration.

The two major types of software that require some form of validation in the test measurement area are data acquisition software and calibration system software. Data acquisition software is designed to obtain electrical signals (voltage, current, frequency) from transducers or signal conditioners, convert the signal into engineering units, carry out calculations, and display and store the results. The validation process in this case involves applying known inputs to the data acquisition system and verifying that the display and storage values are correct. Calculations can be verified by comparing the software results to hand calculations or an alternate computer calculation using previously validated software. The data acquisition timing should also be verified to ensure that data is acquired at the correct intervals.

Validating calibration system software tests the calibration procedures to ensure they operate correctly. For manual entry type procedures this could mean entering known data and verifying that the displayed and stored results are correct. Error calculations and TUR calculations are validated by hand calculations using the same data. Automatic procedures can usually be run in a step mode to verify that the correct calibrator outputs are generated, that the correct results are displayed and stored, and that calculations and pass-fail decisions are correct.

The validation process is similar to a calibration in that a validation procedure is written, the validation is repeated on a periodic basis, and the results of the validation are stored for future reference.

9.10 Quality Assurance System Requirements

A quality assurance (QA) system such as ISO 9000 will require the calibration of all equipment that is used to demonstrate the conformance of a product to specifications. This means that any measurement that is carried out during the production and testing of a component for the purpose of verifying that it meets its design specifications will be performed by an instrument that has been calibrated. This calibration must have been carried out using standards that were traceable to national or international standards.

The QA system will also describe how the calibration records are identified, handled, stored, and distributed. This will include the calibration results, calibration procedures, and calculations.

Another section will describe the steps that are to be followed if an instrument or standard is found to be out of calibration when a recalibration is performed. This is known as a nonconformance and is usually treated by informing the affected parties, reviewing the effects of the nonconformance, reaching an agreement on how to resolve the nonconformance, and carrying out the resolution. The resolution may involve reworking or scrapping affected product or repeating tests on the instrument in question. The causes of the nonconformance should also be determined so that the event does not happen again.

Other areas covered in the QA system will include identification, control, and storage of the measurement equipment. Identification involves attaching labels to the equipment that show the asset number, the calibration date, and the recalibration due date. Other labels may show that the instrument is only used for indication, not measurement, and thus does not require calibration. Control of instruments may also require using labels or devices to seal the access to adjustments in order to show that the calibration has not been tampered with. Storage requirements have to be documented to show that the instrument has not been handled in a way that affects the operation or calibration.

9.11 International, National, and Other Standards

There are several types of standards, which differ in terms of their accuracy and use.

An international standard is one that has been recognized by a number of countries that have formally agreed to use this standard to determine the value of all other standards that define the particular measurement. Most countries now use the SI units of measurement, and all standards except the kilogram are defined in terms of fundamental physical properties. This means that countries now agree on the method for realizing the measurement from the physical properties rather than using some physical object or artifact like the standard kilogram, which was produced in the 1880s and held at the International Bureau of Weights and Measures in France. Copies of this standard kilogram are used by other countries to act as their mass standard, and comparisons with the international standard are made periodically. The length standard, which used to be an artifact, was a bar scribed with two lines a meter apart. This has been replaced by the measurement of how far light travels in a vacuum over a given time period. The time is measured by the atomic clock, which is used to define the value of the second.

A national standard is one that has been recognized formally by the government of a country to act as the value against which all other standards in the country will be compared. Each country usually has a national standards laboratory, which is designated to maintain these standards. Examples of these laboratories include the National Institute of Standards and Technology (NIST) in the United States and the National Research Council (NRC) in Canada.

A primary standard is one that has the best possible metrological qualities for a given measurement so it can be used to calibrate other standards or measurement instruments. The Standard Platinum Resistance Thermometer (SPRT) is a primary standard for temperature calibrations, while a high-precision dead-weight tester is a primary standard for pressure calibrations. Primary standards are maintained and used by a number of different types of organizations. These include government primary standards laboratories, private calibration laboratories, companies that manufacture high level standards, and companies that have an extensive in-house calibration program.

A secondary standard is one that has its value determined directly or indirectly by a primary standard. An organization may hold one standard as its primary standard and only use this standard to calibrate a number of secondary standards. The secondary standards are used on a more frequent basis to calibrate other standards or measurement instrumentation.

A reference standard is usually the best available at one location and acts as the basis from which all measurements made at the location are derived. An organization could have one primary standard and several secondary standards located at manufacturing plants across the country. The secondary standards act as the reference standard for each plant.

A transfer standard is one that is used to compare other standards or to compare a measurement instrument with a standard through a sequential comparison. A high-accuracy digital multimeter can be used as a comparison with a multifunction electrical calibrator, which is then used to calibrate bench-type multimeters. The high-accuracy multimeter in this case acts as a transfer standard since its values are used to calibrate the bench-type multimeters rather than the values directly from the multifunction calibrator.

A traveling standard is one that may have special construction details that allow it to be moved between locations and act as a transfer standard. These construction details could include making the standard more rugged or providing a battery-operated power supply so the standard is powered up during shipment.

A working standard is one that is used to directly calibrate measurement instruments that have a lower accuracy than the standard. An organization could keep one reference standard and have several working standards that are used in daily activities. The working standards could be compared periodically with the reference standard.

An intrinsic standard is one that is based on an inherent physical constant or an inherent or sufficiently stable physical property. Examples include the melting and freezing points of pure metals.

The number and types of standards that an organization uses vary according to how calibration is conducted within the organization. An organization with a single location may not need transfer or traveling standards. Another organization may carry out all its calibrations using primary standards.

9.12 Calibration Equipment Specifications

A large number of factors must be considered when purchasing calibration equipment or comparing equipment from different suppliers. The initial purchase price is only one factor to consider, and it may cost as much or more to maintain the standard over its life.

When considering calibration hardware, the following factors should be evaluated:

What is the range of the equipment?
The calibrator's range should match the ranges of the instruments being calibrated. Too small a calibrator range will mean that the full range of the instrument will not be calibrated. Too large a range may mean that the calibrator is only being used at the lower end of its range

with a resulting lowering of the TUR or TAR. This is the case if the accuracy specification is given as a percentage of full scale rather than as a percentage of reading.

What is the accuracy of the equipment?
The accuracy of the standard should be selected to provide a Test Accuracy Ratio of at least four to one over the full range of values being calibrated. Selecting equipment that would provide a TAR of greater than ten to one is likely not cost effective unless something like anticipated future requirements are a factor. It may be less expensive to purchase two lower accuracy standards to cover a wider range rather than to purchase one high-accuracy standard to provide the necessary TAR.

What is the stability of the equipment?
The stability of the calibrator dictates the length of time that can pass before the equipment has to be recalibrated. This time period is eventually determined from the calibration history of the individual calibrator, which is usually affected by the amount of use, the care in use, and the design. The manufacturer may provide a recommended calibration period or give a stability specification that can be used to calculate a period.

What are the environmental limits and effects?
Each type of standard has specifications that describe the conditions under which it will operate and still maintain a given accuracy. The user must provide a facility that maintains these conditions, including room temperature, humidity, vibration, and cleanliness. If field calibrations are to be conducted, the standard may have to operate in winter to summer conditions and be rugged enough to be used in outdoor plant conditions.

Is a traceable calibration included in the initial cost?
The price of the calibrator may or may not include the cost of a traceable calibration. Some manufacturers may include the cost of a calibration as an option to give the purchaser flexibility in terms of calibration and the calibration supplier. The purchaser should be aware that a certificate of compliance is not the same as a certificate of calibration. The certificate of compliance will merely affirm that the equipment meets

the published specifications for the particular type of equipment. These specifications are based on the statistical data for a large number of the calibrators and not on the calibration results for one particular calibrator.

What is the cost of recalibration?
This cost along with the calibration period will determine the cost of keeping the calibrator calibrated over a long period of time. The cost of the recalibration should include the shipping and packaging costs, which can be expensive due to the relatively fragile construction of some standards.

What is the calibration turnaround time?
The calibration turnaround time determines how long your calibration facility will be unable to provide calibration services for a particular measurement. The facility may have to purchase a second calibrator if the time is excessive. Careful scheduling of recalibrations with the supplier may help to minimize the turnaround time. Having more than one source for recalibration will also be an advantage.

Does the supplier provide the necessary training support?
The operation of sophisticated calibration equipment will require some level of training for the staff that will be using the equipment. The cost of this training and whether the supplier can provide it locally must be considered.

Does the supplier have ongoing technical support?
The supplier should be able to provide technical support for the equipment over its operating life. This can include information about the operation of the equipment, upgrades to the equipment, and provision of spare parts and operating supplies.

Does the supplier have a qualified repair service?
The supplier should maintain a repair facility with qualified staff and equipment in order to carry out any needed repairs to the calibration equipment. This work should be done on a timely basis to minimize the downtime for the user.

Does the supplier have an in-house calibration facility?
The calibration equipment supplier should have the facilities to calibrate the equipment. Ideally, they should be available locally in order to minimize shipping and packaging costs. The equipment purchaser must ensure that the supplier's calibration facility can competently provide traceable calibrations. This may require that the facility be accredited by an authorized body or be audited by the purchaser's quality assurance department.

Is the equipment compatible with the user's standards, auxiliary equipment, and equipment being calibrated?
This generally applies to any fluids that are used by the equipment. The fluid used in a dead-weight tester should be compatible with the fluid being tested or substantial cleaning procedures may have to be employed. Quartz-sheathed SPRTs should not be used in a salt bath because they will become contaminated.

When considering calibration system software, the following should be kept in mind:

Is the software suitable for the user's requirements?
The software may be designed for a specific type of calibration facility such as a mechanical metrology laboratory, a process instrument calibration facility, or an electrical/electronic equipment calibration laboratory. The user should choose a compatible software package or ensure that the package can be adapted to the user's needs.

Does the software have the required flexibility?
Each calibration facility has unique requirements and methods of operation, and the software must have the flexibility to meet these needs. It may be more cost effective to purchase a more expensive flexible software package than to make changes to the way the facility operates. The software's ability to generate reports and procedures are two important areas to consider.

How many assets can the software support?
The system must have room to handle all present and future instruments and standards that could possibly be used by a facility.

Is the database ODBC (Open Data Base Connectivity) compliant?
ODBC compliance allows the user to access any data from the database from any application software regardless of the database management system software being used. This gives the user more flexibility in producing reports and using the calibration data in a number of different ways.

Can the database be backed up easily?
Database backup is extremely important in order to protect the archived calibration data. The backup procedure should be easy to use and flexible enough to meet the user's requirements. Some users may back up to separate disk drives and media while others may use their company's intranet backup facilities.

Does the software enable data to be imported from other databases?
It should be possible to move the information in existing calibration databases to the calibration system database. This should be an automatic procedure once the equivalent fields have been identified.

Can the software produce calibration procedures?
This can involve manual and automatic procedures, the ability to use text and picture files, the ability to control computer interfaces and buses, and the availability of drivers for common calibration equipment. The software should have a built-in inventory of procedures for using typical calibration instruments.

10
Pressure Calibration

Calibrating pressure instrumentation is one of the commonest forms of calibration that laboratories and industrial facilities perform. The calibration equipment for moderately accurate calibrations is relatively inexpensive, rugged, and easy to use. The wide range of pressures that may be measured in one facility will require more than one type or range of calibration standard. The pressure range from one inch of water to 100,000 pounds per square inch, for example, represents a 2.77 million–to–one ratio of maximum pressure to minimum pressure. The number of standards needed is also increased if vacuum and absolute pressure calibrations are required and if various fluids such as water, oil, and gas are required.

Pressure is a derived unit based on the fundamental units of length (l), mass (m), and time (t). Earlier chapters of this book defined pressure as a force acting on an area (l^2), defined force as a mass (m) being accelerated, and defined acceleration as (l/t^2). The accuracy of many of the pressure standards depends on how accurately we can produce, maintain, and measure these fundamental units when manufacturing and using the standards.

10.1 Dead-Weight Testers

Dead-weight testers (see Figures 10-1, 10-2, and 10-3) are the most widely used type of pressure calibrator and are available in a variety of accuracies, fluid types, ranges, and pressure units. A typical dead-weight tester consists of a fluid-filled chamber that is pressurized by an integral hand pump, a ram screw, or an external source of pressure. A vertical cylinder together with a close-fitting piston that rotates within the cylinder is built into the chamber wall so the fluid pressure acts on the bottom area of the piston. A pressure port in the chamber is provided so the pressure instrument being calibrated can be connected. The fluid pressure acting on the bottom area of the piston produces an upward force that lifts the piston. The upward force is balanced by the downward force produced by the addition of weights to the top of the piston. This force (known as the weight) is equal to the mass of the weights plus the mass of the cylinder times the acceleration due to gravity. When the two forces are equal, the piston will float within the cylinder.

Figure 10-1. Dead-Weight Tester
(Courtesy, GE Sensing)

Figure 10-2. High Line Pressure Differential Dead Weight Tester
(Courtesy, GE Sensing)

Figure 10-3. Primary Standard Dead Weight Tester
(Courtesy, GE Sensing)

Upward force = pressure × piston area

Downward force = mass × acceleration due to gravity = weight

At balance (piston floating):

Upward force = Downward force

Pressure × piston area = weight

Pressure = weight / piston area

Piston and cylinder assemblies are manufactured with extremely small annular gaps between the two parts. The fluid fills this gap and acts as a lubricating film between the two surfaces. This film reduces wear and friction between the two parts, and any axial viscous friction caused by the film is reduced by slowly spinning the weights and piston at the balance point. A tiny amount of the fluid will eventually escape past the piston, and the piston will slowly sink within the cylinder. This leakage is overcome by using a trimming ram screw that provides a small increase in pressure to raise the piston again.

The basic accuracy of a dead-weight tester is determined by how accurately the piston area and mass of the weights can be measured. The accuracy of testers ranges from 0.1% of reading for basic working standards to 0.003% of reading for a primary standard. Reaching the primary standard accuracy requires compensating for a large number of factors such as local gravity, air buoyancy, temperature, air density, fluid surface tension, and piston/cylinder deformation under high pressure. Calculating these effects is usually done by a computer program supplied by the tester manufacturer. The major effect that should be accounted for in lower accuracy testers is the variation in gravity since this directly affects the weight value of the masses. The value of gravity can vary as much as 0.17% over the continental United States and depends on the latitude, height above sea level, and presence of abnormal rock densities below the ground. Estimates of local gravity can be obtained from government sources, but a standards laboratory will have a precise measurement of the gravity made at the location of the dead-weight tester. The manufacturer can supply weights that are adjusted for the local gravity. Otherwise, they are supplied for a standard gravity of 980.665 cm/sec^2, and the user would have to multiply the nominal weight value by (Local Gravity / 980.665) to obtain the effective weight.

The pressure range covered by dead-weight testers is from approximately 5 psi to 100,000 psi, although it would require more than one tester to cover the entire range. The minimum pressure is determined by the weight of the piston and weight carrier assembly since this is the minimum weight that can be floated. Absolute pressure calibrations are conducted using a tester that has the weights and piston in an enclosure that is evacuated to as close as possible to a perfect vacuum. Vacuum calibrations are carried out with a tester that is essentially built upside down and uses a variable vacuum source to lift the piston and the suspended weights to a floating position.

A typical working standard dead-weight tester might have two interchangeable piston and cylinder assemblies with different piston areas and a range of weights that would allow the tester to cover the range from 5 to 10,000 psi.

The piston areas would be 0.1 square inch for the low-range operation and 0.01 square inch for the high-range operation. Each piston and weight carrier would have a weight of 0.5 pounds.

A weight set would be supplied as follows:

- 4 weights at 0.5 pounds
- 4 weights at 2.0 pounds
- 1 weight at 9.5 pounds
- 9 weights at 10.0 pounds

Using the 0.1-square-inch piston the minimum pressure would be obtained by floating just the piston and weight carrier, which weigh 0.5 pounds.

The resulting pressure would be 0.5 / 0.1 = 5 psi.

Adding one 0.5 pound weight to the weight carrier would increase the total weight to 1.0 pounds, and the resulting pressure at balance would be 1.0 / 0.1 = 10 psi. Additional 0.5 pound weights result in an increase of 5 psi per weight.

Adding 2.0 pound weights increases pressure by 2.0 / 0.1 = 20 psi per weight.

The 9.5 pound weight is supplied for convenience since the weight plus the piston and carrier weigh 9.5 + 0.5 = 10.0 pounds. This combination results in a pressure of 10.0 / 0.1 = 100 psi. Each additional 10.0 pound weight will increase the pressure by 10.0 / 0.1 = 100 psi.

The total weight of all the weights plus the piston and weight carrier is 110.0 pounds. This combination would result in a pressure of 110.0 / 0.1 = 1,100 psi.

Using this set of weights and the 0.1-square-inch piston, any pressure between 5 psi and 1,100 psi in increments of 5 psi can be produced.

If the 0.01-square-inch piston is used, the minimum pressure produced by floating just the piston and weight carrier is 0.5 / 0.01 = 50 psi. Adding weights as previously described will produce pressures that are 10 times higher than those produced by the 0.1-square-inch piston. The same set of weights and the 0.01-square-inch piston will produce any pressure between 50 and 11,000 psi in 50 psi increments.

Some manufacturers can supply a second set of weights that will produce pressures in different units while still using the same piston and cylinder assemblies. This requires that a special weight be added that will always stay on the weight carrier to produce a meaningful minimum pressure in the new units. For the tester previously described, the 0.1-square-inch piston and weight carrier weighed 0.5 pounds and produced a pressure of 5.0 psi. Converting the tester to produce units of kilopascals would require the following:

0.5 pounds produced 5.0 psi, which is equal to 34.47 kilopascals, which is close to 50 kilopascals.

50 kilopascals is equal to 7.2519 psi, which would be produced by a weight of 0.72519 pounds on the 0.1-square-inch piston. This would be done by producing a special weight of (0.72519 − 0.5) = 0.22519 pounds, which would always stay on the weight carrier.

The weight set would be produced as follows:

- 4 weights at 0.72519 pounds, producing 50 kilopascals per weight
- 4 weights at 2.90076 pounds, producing 200 kilopascals per weight
- 1 weight at 13.7786 pounds, producing 950 kilopascals per weight
- 9 weights at 14.5038 pounds, producing 1,000 kilopascals per weight

This set would produce any pressure from 50 to 11,000 kilopascals in 50 kilopascal increments. The same set of weights used with the 0.01-square-inch piston would produce any weight from 500 to 110,000 kilopascals in increments of 500 kilopascals. This may be a less expensive option compared to purchasing two separate dead-weight testers.

The fact that the dead-weight tester applies pressure in increments can be a problem in certain calibration situations. If the standard calibration procedure instructs the user to calibrate the instrument at ten equally spaced steps over its range, then there may be certain ranges that cannot be done this way. The tester described previously could not calibrate a transducer with a range of 1600 psi at ten equally spaced steps of 160 psi since the minimum step over 1100 psi is 50 psi. A separate procedure that allowed a calibration using eight equally spaced steps would have to be used. The incremental pressure steps also make the dead-weight tester difficult or impossible to use when determining transducer characteristics such as hysteresis and linearity since the pressure is increasing and decreasing as the weights are added and removed. Overshooting the calibration points is also common unless great care is used when adding weights or increasing pressure.

Other considerations when using dead-weight testers include making sure that the tester is leveled and eliminating or accounting for errors produced by locating the instrument being calibrated above or below the reference height of the tester. The height of liquid will cause an error equal to the height times the density of the fluid. The reference level of most dead-weight testers is at a plane through the center of the vertical travel of the piston.

The main concern in the maintenance of dead-weight testers is keeping the tester free of dirt and debris. External dirt can change the mass of weights, while dirt and debris in the tester fluid can prevent internal check valves from closing, plug small orifices, and even enter the small clearance between the piston and cylinder. This can cause binding or surface scoring and, eventually, cause an expensive component to be scrapped. The instrument to be calibrated should be cleaned as much as possible before it is attached to the tester for calibration. Filters are available for some testers to reduce the introduction of dirt and debris into the tester fluid.

10.2 Water Dead-Weight Testers

A water dead-weight tester is one that uses water as the fluid within the body of the tester. The water is pressurized and acts on the piston and the instrument being calibrated. The testers are usually manufactured with materials that are compatible with distilled water so that instruments used in the food industry can be calibrated. The use of distilled water also prevents contamination of water-based hydraulic systems and other systems that would be affected by the oil used in oil dead-weight testers. Contamination by oil of components that are used in oxygen service is also eliminated by using water as a pressure medium. Oil coming into contact with high-pressure oxygen can result in severe explosions.

The disadvantage of water is the fact that it is not a very good lubricator, and it will evaporate at room temperatures. This results in poorer piston-spinning action, dry contact between piston and cylinder, and more leakages from valves and pumps.

10.3 Oil Dead-Weight Testers

Oil-filled dead-weight testers are the most common type used in working standard level and certainly primary standard level calibrators. The lubricating qualities of oil compared to water provide a much better medium for the mechanical components used in the testers. Some manufacturers can supply diaphragm assemblies that isolate the tester oil from the fluid that is exposed to the instrument being calibrated. This also requires the use of a second pressurizing system and a means for indicating equal pressure on both sides of the diaphragm. This should only be considered if no other options are available.

10.4 Pneumatic Dead-Weight Testers

The pneumatic dead-weight tester (see Figure 10-4) uses pressurized air or other gases as the pressure medium.

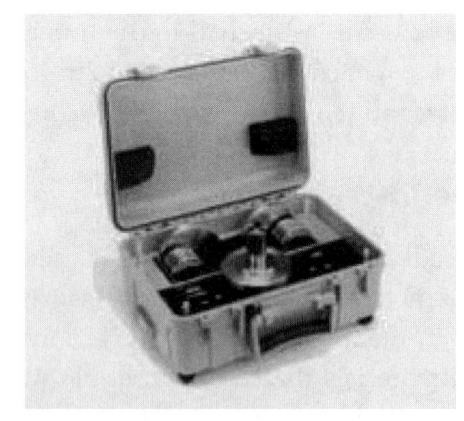

Figure 10-4. Pneumatic Dead Weight Tester
(Courtesy, GE Sensing)

The construction of a pneumatic dead-weight tester is different than that of water or oil tester, but it still operates on the principal of an upward force of pressure acting on a known area, which is balanced by the downward force of weight. The pneumatic tester uses a ceramic ball that is located within a tapered nozzle to support a weight carrier. The gas acts on the underside of the ball and lifts it and the weights in the tapered tube until the flow past the ball is equal to the supply flow that is controlled by a flow regulator. At this point, the system is in equilibrium, and the pressure that is required to maintain the flow (and ball position) is proportional to the weight. The system is self-balancing in that added weight causes the ball to sink and restrict the flow in the ball/cylinder annulus. The flow regulator will increase the pressure until the ball rises to the position that maintains the set flow. A reduction in weights will cause the opposite action to occur. At balance, the ball is centered in a film of air that eliminates contact and friction between the ball and nozzle.

The advantages of the pneumatic tester include the self-balancing action that eliminates the need for any pumps and the lack of contamination from oil or water. The use of a gas as the pressurizing fluid almost entirely eliminates errors caused by any height difference between the instrument being calibrated and the reference level of the tester. The disadvantages of the pneumatic tester include limited ranges and the need for a supply of clean, dry gas at a pressure greater than the upper limit of the tester range. This can be provided by a compressed instrument air system or by high-pressure bottled gas. Either supply usually limits the portability of the pneumatic tester.

10.5 Digital Calibrators

Pressure standards are also available that use highly accurate and stable pressure transducers as the reference elements. These standards usually use microcomputer-controlled digital displays to show the pressure reading in various units (see Figure 10-5). Models are available that use liquids or gases as the pressure medium, and some units have internal servo-controlled pressure systems that automatically produce a programmed series of pressure steps.

The most accurate versions of these calibrators use quartz sensing elements that are highly stable, highly elastic, and have low hysteresis

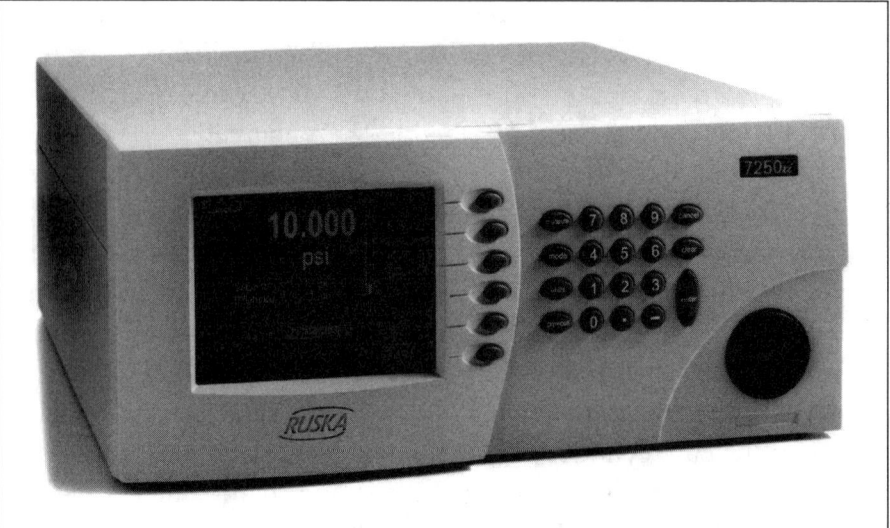

Figure 10-5. Digital Pressure Calibrator
(Courtesy, GE Sensing)

and high repeatability characteristics. Some units use fused quartz and are made in the form of conventional pressure elements such as bourdon tubes and helixes. Other resonant quartz units use standard material pressure elements such as bellows or bourdon tubes to impart a force onto an oscillating quartz crystal. The frequency of oscillation is proportional to the applied force in a highly reproducible manner. Temperature compensation is provided by using a second oscillating crystal that is subject to the same temperature but not to the pressure so as to bias out the frequency changes with temperature. These calibrators can be provided in a wide variety of pressure ranges from a few inches of water to 40,000 psi. They have typical accuracies that are equal to or better than 0.01% of full scale pressure with stabilities within 0.01% of full scale per year.

Lower accuracy calibrators are made using conventional piezoresistive or resonating silicon pressure transducers. These calibrators are typically working level standards and are often packaged for portable use in outdoor or indoor plant conditions (see Figure 10-6). The accuracy of these units is typically 0.025% to 0.05% of reading and pressure ranges from a few inches of water to several thousand psi. Some units use a series of plug-in pressure modules, each covering a particular pressure range. This allows the manufacturer to maximize the accuracy for each specific application rather than relying on the linearity of

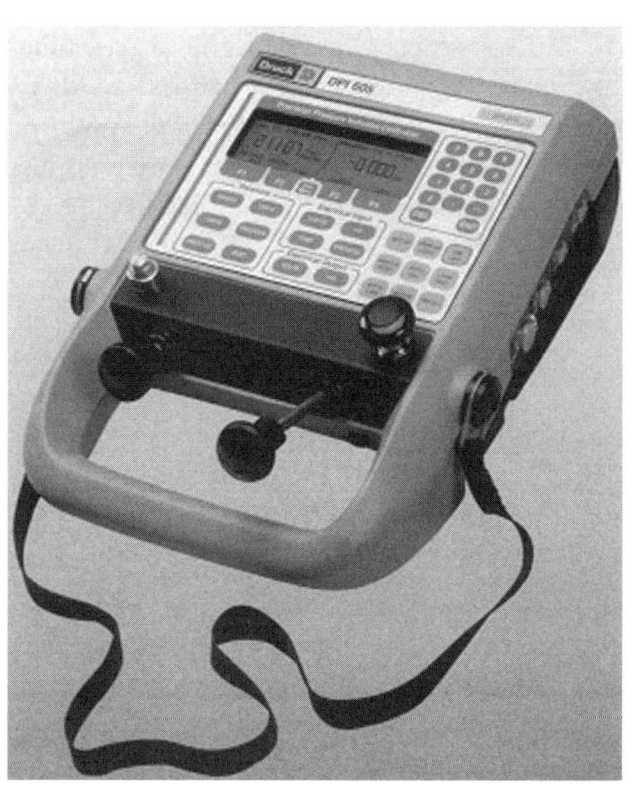

Figure 10-6. Portable Digital Pressure Calibrator
(Courtesy, GE Sensing)

a single transducer to provide high accuracy over a wide pressure range. These types of calibrators are often supplied with integral pumps, transducer power supplies, digital voltmeters, barometers, and microcomputer-controlled display and storage capabilities. This allows the unit to function as a complete calibration facility with the ability to generate pressure, power transducers, display in multiple units of absolute or gauge pressure, and store and download calibration results.

10.6 Manometers

In theory, a manometer can be considered as a primary standard since it measures pressure by using the fundamental units of length and mass. In practice, there are many factors that limit its use as a practical calibration tool.

The simplest form of a manometer is the U-tube manometer (see Figure 10-7). This is constructed by forming glass tubing in the shape of a 'U' and partially filling the tube with a liquid. With equal pressure applied to each leg, the height of the liquid in the one leg will be equal to the height in the other leg. If a pressure is applied to the right-hand leg then the height of the liquid will fall in the right leg and rise in the left leg. The pressure is equal to the ratio of the force applied to the surface area of the liquid in the tube:

$$\text{Pressure} = \text{Force} / \text{Area}$$

Applying more pressure will cause the level in the right leg to fall further and the level in the left leg to rise further. Further changes in pressure show us that the difference in height between the two legs is proportional to the applied pressure.

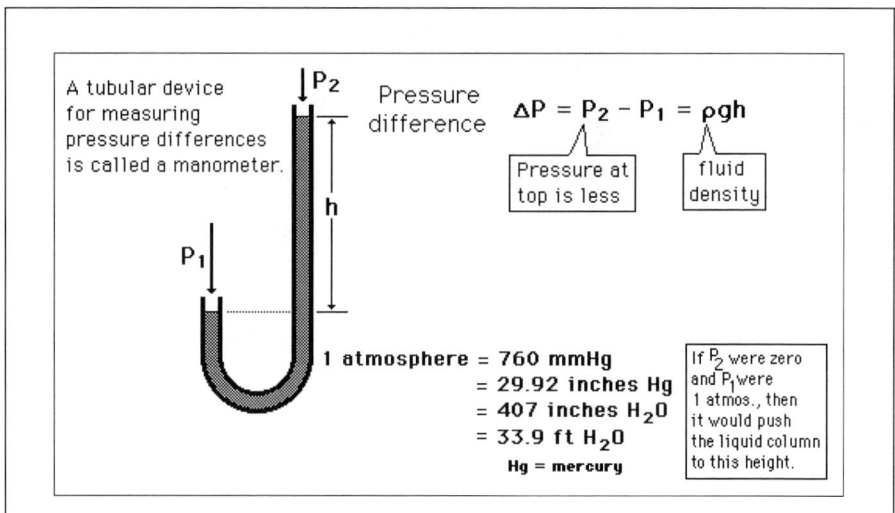

Figure 10-7. U-tube Manometer

This difference in height of the liquid causes a force equal to the mass of liquid times the acceleration due to gravity, which equals the force created by the pressure being exerted on the fluid area:

$$F = m \times g$$

This force is distributed across the cross-sectional area of the tube, producing a pressure that is equal to the applied pressure:

$$\text{Pressure} = (m \times g) / \text{Area}$$

The mass of liquid is equal to the volume times the density, and the volume in the tubing is equal to the cross-sectional area times the height.

This results in:

$$\text{Pressure} = (\text{density} \times \text{Area} \times \text{height} \times g) / \text{Area}$$

or

$$\text{Pressure} = (\text{density} \times \text{height} \times g)$$

This equation shows that the maximum pressure that can be read by a manometer with a reasonable length is limited by the density of the fluid. Using mercury as the fluid, a 100-inch manometer can only measure approximately 49 psi. Using water as the fluid, a 100-inch manometer can measure approximately 3.6 psi. This relatively narrow range is one of the disadvantages of manometers.

The fact that the pressure reading is proportional to density means that the operator must make corrections for density changes with temperature. Temperature can also change the length of the scale that is used to measure the fluid height. Other disadvantages include the difficulty in reading the fluid height as a result of parallax errors and difficulty in keeping the internal tube walls clean. Dirty walls will cause variations in the meniscus height and will contaminate the fluid and change its density. These disadvantages limit the practical application of the manometer as a pressure standard. Low-cost digital manometers that use transducers to measure pressure have largely replaced the liquid manometer.

11
Temperature Calibration

To calibrate temperature sensors you need a source of heat and a calibration standard. The sensor being calibrated and the standard are compared as they are exposed to the same temperature conditions provided by the heat source. To calibrate temperature signal conditioning or displays you need a standard that simulates the sensor output. Alternatively, the sensor and conditioner or display can be calibrated as a system by comparing the system output to the temperature measured by a standard that is exposed to the same temperature as the system sensor.

The overall cost of performing temperature calibrations is usually higher than for other types of calibrations. This is because of the cost of calibration equipment, the cost of calibration facilities, and the amount of time spent on the calibration. The calibration time is dictated by the time it takes the heat source to reach a set temperature and stabilize at that temperature. This may take close to one hour when using precision temperature baths, and the only way to reduce time is to provide multiple baths that are set at different temperatures. In comparison, pressure changes using dead-weight testers can be accomplished in less than a minute. The facilities required to support temperature calibration include extraction fans or fume hoods, increased air conditioning capacity, and high levels of electrical power. These facilities and the size and weight of typical calibration baths means that the sensor will have to be removed from its location to a calibration laboratory for calibration. There are some portable temperature calibration units that can be used in the field for moderate accuracy calibrations. These include dry block calibrators and mini-baths.

Safety is an issue that must be taken seriously when performing temperature calibrations. The operator is dealing with large volumes of very hot or very cold liquids that can cause bodily harm or damage other equipment. Proper training is required to ensure that operators understand and deal with the potential dangers associated with temperature calibration.

11.1 Temperature Calibration Baths

A calibration bath (see Figures 11-1, 11-2, and 11-3) is the most common form of heat source used to calibrate temperature sensors. A good temperature bath consists of a lot more than just a container of fluid and an external or internal electrical heater. The most important specifications for the bath include temperature stability and temperature uniformity. Both of these specifications should be at least ten times better than the required uncertainty of the device being calibrated. For calibrations with an uncertainty of ± 0.1° Fahrenheit this means that the bath temperature control system must maintain the fluid temperature to within ± 0.01° Fahrenheit. The same degree of control is required over the complete calibration range of the bath. This level of control should be maintained at least for a short-term period of 15 to 20 minutes for the system and sensors to equalize and the temperature readings to be taken. Longer-term stability is an asset if the bath is maintained at a fixed temperature for hours or days at a time. The temperature uncertainty specification means that the bath must also maintain the fluid temperature to within ± 0.01° Fahrenheit everywhere within a certain fluid volume, which is designated by the manufacturer as the calibration zone. This ensures that sensors and standards placed within the zone are at the same temperature within ± 0.01° Fahrenheit. The stability and uniformity specifications obviously have to be better to achieve more accurate calibrations.

The bath's stability is affected by the heater control system, the characteristics of the fluid, and the temperature being controlled. A simple on-off temperature control system will not likely produce sufficient stability. Some form of proportional and integral control functions are necessary to maintain tight control and eliminate offsets from the

Figure 11-1. Temperature Calibration Baths
(Courtesy, Hart Scientific)

Figure 11-2. Mini Temperature Calibration Baths
(Courtesy, Hart Scientific)

required controller set point. The viscosity of the fluid affects stability, and the user should know what fluid was used to produce the stability specification. Stability is also affected by the heat loss or gain to the fluid from the ambient surroundings. As the control point is moved away from room temperature the bath's insulation must be good enough to minimize the heat losses or gains and maintain stability.

The bath's uniformity is largely controlled by the mixing of the fluid. The mixing must prevent temperature gradients from occurring in the bath's calibration zone. This means that horizontal and vertical gradients as well as circulating flow patterns must be avoided. Mixing is usually provided by circulating pumps or motor-driven stirrers. Circulating pumps tend to produce definite flow patterns from outlet to inlet and may not spread the flow over the whole zone. Some stirrers will only move the fluid in a circular direction, which provides good vertical but poor horizontal uniformity. This can be a problem if sensors with different immersion lengths are calibrated at the same time. Propeller-style stirrers that move the liquid in both directions provide the best bath uniformity.

Figure 11-3. Micro Temperature Calibration Baths
(Courtesy, Hart Scientific)

A good bath also includes other features, such as an adjustable upper temperature trip device, auxiliary start-up heaters, extra bath lids that can be drilled for sensor access, and good outer surface insulation or cooling to protect personnel. A high-resolution digital display of the controlled temperature is needed to show when a stable temperature is reached. The actual bath temperature is measured with a SPRT or high-accuracy secondary standard RTD.

The temperature range covered by liquid temperature baths in general is approximately -100°C to 550°C. This range requires the use of several different fluids, ranging from low-temperature heat transfer fluids to oils and molten salt. No one fluid will work over the entire range. The limitations of various fluids include freezing points, boiling points, flash points, and high viscosity at lower temperatures. High viscosity will make circulating and mixing difficult because of the high power levels required to move the fluid.

11.1.1 Oil Baths

Oil baths typically use silicon oils, mineral oil, or low-temperature fluids to cover the range from -100°C to approximately 300°C. The low-temperature fluids range from -100 to 100°C, while the oils range from -40 to 300°C. Oils that are good at the lower temperature will usually have a lower flash point that limits their use at higher temperatures. The flash point is the temperature at which the vapor (not the fluid) will ignite if exposed to an open flame. The vapor will stop burning if the flame is removed. Oils that are good at high temperatures tend to have too high a viscosity to be used at lower temperatures. The bath should have a temperature trip device that prevents the circulating device from operating if the temperature is low enough to produce damaging viscosity levels.

If calibrations over a wide range are required, it makes economic sense to have a number of baths that use different fluids. These baths can be set at different temperatures to speed up the calibration process. The sensor only has to be cleaned and moved from bath to bath. The alternative would be to change the fluids used in one bath. This would require excessive time to cool and drain fluid, clean the bath, and reach a new temperature equilibrium using a new fluid.

Oil baths operated at high temperatures will produce smoke fumes that have a disagreeable smell. A fume hood or other form of air extraction equipment is required when using these fluids.

11.1.2 Water Baths

Baths that use water as the fluid medium are designed in much the same way as oil baths. The limited temperature range and low fluid viscosity lower the cost of the bath because the heater is smaller, less insulation is required, and mixing power needed is smaller. The range of a water bath is between 0 and 100°C. The use of a 50/50 mixture of ethylene glycol and water allows a range from -30°C to 90°C. The advantages of using water include the elimination of sensor cleaning after calibration and the elimination of fumes and potential fire hazards.

11.1.3 Sand Baths (Fluidized Bed)

A sand bath or fluidized bed bath consists of a heated volume of fine-grain aluminum oxide. The aluminum oxide has the appearance and consistency of sand. A low-pressure, high flow of dry inert gas or air is applied evenly to the bottom area of the aluminum oxide. This air flows upward through the particles and lifts and suspends the whole bed in a state similar to boiling liquid. The constant agitation of the particles provides a fairly uniform temperature volume into which sensors and reference probes can be easily inserted.

The stability and uniformity of a fluidized bath is not nearly as good as a liquid bath. A special technique called "dead bed operation" is used to partially overcome this limitation. The sensor being calibrated and the reference RTD are inserted into close fitting holes in a block of metal. This block is known as an equalizing block or isothermal block. The bath and block are heated to the desired temperature, and then the heaters are shut off, and half a minute later the air is shut off. This causes the fluidized bed to collapse and produce a dense insulating volume around the block. The temperature of the block remains stable for several minutes, during which the readings from the sensors can be taken. Even this technique does not produce the stability and uniformity levels provided by a good liquid bath.

The advantages of fluidized baths include wide temperature range (50°C to 700°C), the need to use only one bath fill material, no danger of fluid or vapor fire, no fumes, and no postcalibration probe cleaning. Their disadvantages include less consistent stability and uniformity, less precise temperature control, and the difficulties entailed in using equalizing blocks. Equalizing blocks have to be prepared for each different probe diameter, and the block surfaces tend to oxidize at high temperatures. This oxide can reduce the hole diameter or even cause a probe to seize within a hole. The flow of air through the bed carries a fine dust that is difficult to keep within the bath container. A fume hood or air extraction unit may be required to remove the dust and hot air from the room.

11.1.4 Salt Baths

Salt baths use molten salt as the temperature medium. The salt is typically a mixture of potassium and sodium nitrates or nitrites and has a melting point of approximately 150°C. This allows a salt bath to cover the range from 180°C to 550°C. The viscosity of molten salt is very close to that of water, which allows the bath to produce very high levels of stability and uniformity. Other advantages include the high upper temperature limit and the lack of fire danger associated with oil use. Disadvantages include the fact that bath salt solidifies below 150°C, the probe must be cleaned after calibration, and odors are produced at high salt temperature. Fume hood or air extraction units will likely be required when using a salt bath.

11.2 Dry Block Calibrators

Dry block or dry well calibrators (see Figure 11-4) work much like the equalizing block in a fluidized bath. The calibrators have a fairly large block of metal in which deep holes with a diameter only slightly larger than the probe diameter are drilled. This block is maintained at a set temperature using electrical heaters that surround the block. A temperature sensor is located in the block and used for the heater control and the digital display of the block's temperature. Units are available that have various temperature ranges from -45°C to 650°C. The stability and uniformity of dry-well calibrators designed for field use are not as good as fluid or sand baths. Typical stabilities range from ±0.02°C to ±0.05°C, while uniformity from one hole to another is approximately ±0.1°C. This limits the use of dry block calibrators to the calibration of industrial temperature sensors, such as RTDs, thermocouples, and thermistors. Some manufacturers also make dry well calibrators for use in a calibration laboratory that have improved set point accuracy and stability and uniformity specifications that are significantly better than the field calibrators. The stability of these units can be ± 0.01°C or better, while the hole-to-hole uniformity can be ± 0.05°C or better.

Calibrators are available with removable insert blocks that are drilled with one or more holes to suit different probe diameters. In many

Figure 11-4. Dry Block Calibrators
(Courtesy, Hart Scientific)

cases, one hole is used for the probe being calibrated, while another hole is used for a high-accuracy secondary standard RTD. This can provide a more accurate temperature measurement of the block than is possible with the sensor used to control the calibrator. The use of a secondary standard RTD is more important for removable insert blocks that are placed into the larger block that contains the control sensor.

The holes in the wells or blocks are typically 5 to 6 inches deep in order to minimize sensor errors due to stem conduction. Shorter probes should be calibrated by raising the secondary standard probe to the same height as the short probes in order to minimize vertical temperature gradient effects.

11.3 Temperature Calibration Furnaces

Temperature calibration furnaces (see Figure 11-5) are typically designed for the high-accuracy calibration of sheathed thermocouples used at temperatures up to 1,100°C. The calibration range of the furnaces is from 300°C to 1,100°C, although lower temperature operation is possible. Lower temperature calibrations are usually carried out in oil or sand baths. The furnaces use long, removable isothermal blocks drilled with multiple holes that allow immersion depths of 10 to 16 inches. A SPRT or standard S or R type thermocouple is also inserted into the block for temperature comparison purposes. The blocks are usually mounted horizontally to make it possible to insert long thermocouples and to minimize vertical temperature gradients.

Figure 11-5. Temperature Calibration Furnace
(Courtesy, Hart Scientific)

The stability of a furnace ranges from 0.05°C at 300°C to 0.1°C at 1,100°. The hole-to-hole uniformity ranges from 0.05°C at low temperatures to 0.3°C at high temperatures. The large block sizes result in heat-up, cool-down, and stabilization times that can each last several hours. Heater electrical powers in the range of 3,000 to 4,000 watts are required for these furnaces. This typically means a supply voltage requirement of 230 volts AC rather than the commonly available 115 volts AC.

11.4 Infrared Calibrators

Infrared calibrators (see Figure 11-6) are used to calibrate hand-held infrared pyrometers or industrial infrared sensors. The calibrator consists of a heated block with a surface that provides a black body target with a typical emissivity of 0.95. This emissivity matches the value of most fixed emissivity sensors, while variable emissivity sensors or pyrometers can be set at this value. The block may be provided with a well in which a reference sensor can be inserted to increase the calibration accuracy.

Figure 11-6. Infrared Temperature Calibrator
(Courtesy, Hart Scientific)

High-temperature calibrators cover the range from 50°C to 500°C, while low-temperature units with built-in coolers range from -30°C to 200°C. Typical accuracies using the built-in digital temperature displays ranges from ±0.5°C to ± 1.0°C.

11.5 Thermocouple and RTD Simulators

To calibrate signal conditioning equipment or temperature displays you must use devices that simulate the output of the temperature sensors. These devices include RTD simulators that provide a resistance output and thermocouple simulators that provide the DC millivolt signal for the various types of thermocouples.

RTD simulators (see Figure 11-7) are highly accurate, stable decade resistance boxes that are designed to cover the resistance ranges of common RTDs. Calibrators that are designed to simulate 100 ohm RTDs would cover the resistance range from 10 to 1,111 ohms with a resolution of 0.001 ohms and an accuracy of up to ± 0.005 ohms. For a 100 ohm platinum RTD this represents a temperature resolution of 0.0025°C and an accuracy of ± 0.0125°C.

Thermocouple simulators are necessary in order to provide a stable, noise-free DC millivolt level signal for the calibration of temperature displays and signal conditioners. Today's simulators are microprocessor controlled because of the variety of factors that must be addressed

Figure 11-7. RTD Simulators
(Courtesy, Hart Scientific)

in producing an accurate calibration output signal. These factors include:

- Each type of thermocouple has a unique output-versus-temperature curve.

- The cold junction temperature has to be measured and accounted for on a real-time basis.

- The output range is small (0 to 42 millivolts for a J type thermocouple)

- The output resolution requirement is high (40 to 50 microvolts per degree for a J type thermocouple)

- Fahrenheit and Celsius readings must be displayed.

- The required output can usually be selected by pushbutton.

The accuracy of typical simulators for field calibrations is ± 0.1°C. The accuracy of laboratory type calibrators is approximately ± 0.01°C.

11.6 Standard Platinum Resistance Thermometers

Standard Platinum Resistance Thermometers or SPRTs (see Figure 11-8) are high-accuracy, high-stability RTDs that are used to interpolate between the temperatures of the various metal melting and freezing points that are used to define the International Temperature Scale of 1990 or ITS-90. The temperature range of SPRTs is from -200°C to approximately 1000°C. The 0°C resistance (R_0) of SPRTs that are used to cover the range from -200°C to 500°C is usually 25.5 ohms. The 0°C resistance of SPRTs that are used to cover the range from 0°C to 1000°C is usually 2.5 ohms.

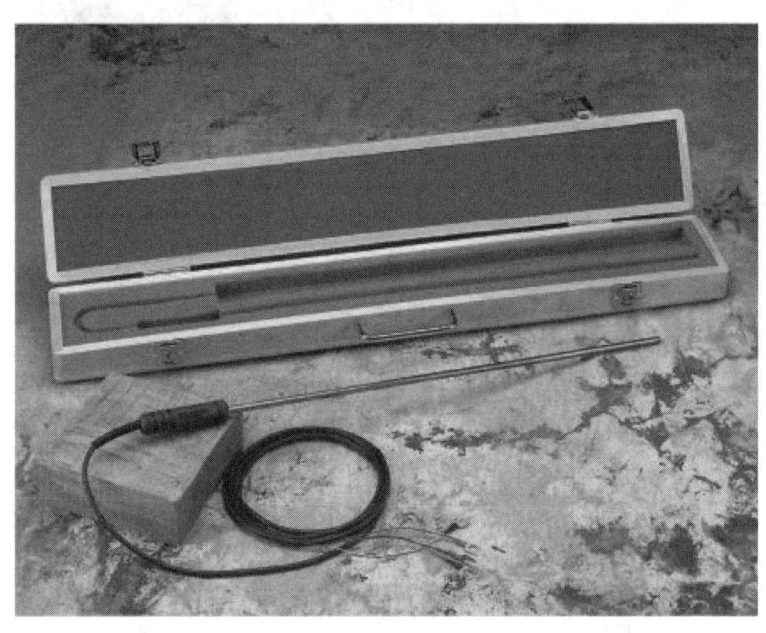

Figure 11-8. Standard Platinum Resistance Thermometer
(Courtesy, Hart Scientific)

The sensing element of these standards is made of ultra-pure platinum wire that is wound on a support in a manner that eliminates any temperature-induced strains in the wire. Any such temperature-induced strain will contribute a resistance change in the element in addition to the resistance change due to temperature alone. This contribution will cause an error in the output signal of the SPRT. The SPRT is typically configured as an 18- to 24-inch long probe with a stainless steel or quartz sheath whose diameter is approximately one-quarter inch. The

SPRT is supplied with a six- to eight-foot cable containing four wires that are terminated with gold-plated spade lugs. The readout device (see Figure 11-9) is a high-accuracy resistance bridge with an input circuit that uses the four wires to cancel any lead wire resistance effects. The long probe length enables the user to insert the SPRT into the bath fluid with a minimum immersion depth of eight inches while still keeping the handle and cable cool. This minimum depth prevents stem conduction errors from affecting the measurement. The long cable length ensures that the resistance bridge or other readout can be located far enough from the bath to prevent ambient temperature effects.

Figure 11-9. SPRT Precision Thermometers
(Courtesy, Hart Scientific)

The SPRT is a fragile instrument that must be handled with great care. Typical primary standard laboratory calibrations are carried out to uncertainties of less than 1 mK (one thousandths of a degree Kelvin). A SPRT that is tapped slightly on a hard surface can suffer a change in resistance that results in a temperature measurement change of 1 to 2 mK. Using a SPRT at temperatures above 500°C requires other precautions. At high temperatures, contamination of the platinum wire from metal ions in the sheath or in metal contacting the sheath is a real possibility. The cool-down rate from temperatures above 500°C should be limited to about 2° per minute to avoid thermal shocks that can change the resistance. The effects of thermal and mechanical shocks can possibly be reversed by annealing the SPRT. This annealing should maintain the SPRT at a temperature slightly higher than the maximum use temperature for approximately 12 hours. The SPRT should be cleaned of any bath fluids or other contaminants as soon as possible using pure ethyl alcohol. Such cleaning should also take place before using the SPRT at high temperatures in order to remove any contamination to the base metal surface. Quartz-sheathed SPRTs should not be used in salt baths unless a thin, close-fitting metal or pure ceramic protective sheath is used over the quartz sheath.

Many adverse effects on SPRTs show up as a change to R_0 or the resistance at 0°C. These potential effects should be monitored by frequently checking the R_0 using a well-made ice bath or a triple point of water cell.

11.7 Triple Point of Water Cells

A triple point of water cell (TPW) is probably the most useful and least expensive primary standard that exists (see Figure 11-10). A TPW can be purchased for approximately $1,000, and support equipment will add between $500 and $3,000, depending on the user's work load for the TPW.

The temperature realized by a TPW is a defining point on the International Temperature Scale of 1990 (ITS-90). It is assigned the value of 0.01°C or 273.16 K. The TPW is considered an intrinsic standard, and the value is determined by the naturally occurring physical properties of the water in the cell. The triple point refers to the fact that when the solid (ice), liquid (water), and gaseous (water vapor) phases of pure, airless water occur together in equilibrium, then the temperature at the solid-liquid interface will be 0.01°C. The TPW cannot be calibrated since by definition it is the most accurate representation of the temperature point, and any other temperature measuring device is not as accurate. The manufacturer will qualify and certify that the TPW meets the requirements for obtaining the equilibrium temperature by comparing its operation with a number of other TPWs maintained for that purpose. A detailed history of the use and operation of this group of TPWs should be maintained as evidence of their suitability for use as comparison cells. Unless physical damage occurs, the life of a TPW is 10 to 20 years.

Figure 11-10. Triple Point of Water Cell
(Courtesy, Hart Scientific)

The triple point of water cell consists of a sealed cylinder made of borosilicate glass with a co-axial inner tube that is open to the atmosphere at the top. This tube or re-entrant well is used to hold the temperature probe being calibrated. The bottom, sealed end of the re-entrant well is not as deep as the outer cylinder in order to allow water to completely surround the well. The sealed volume between the outer cylinder and the re-entrant well is almost completely filled with ultra pure, air-free water.

The TPW is prepared by causing a layer of ice to form around the sides and bottom of the re-entrant well. This solid layer of ice is known as a mantle. The mantle must not bridge the complete space between the well and the outer cylinder since the resulting forces could break the glass. Once the mantle is formed, a room-temperature rod is inserted into the well and allowed to slightly heat the glass wall. A spinning motion of the cell about its long axis is then performed to loosen the ice mantle from the well wall. This action allows a thin film of water to exist between the wall and the ice. The cell now has the ice and water phases along with the water vapor–filled space above the water to provide the triple point environment and the resultant triple point temperature at the ice-water interface. The prepared cell is completely immersed in an ice bath to maintain the conditions over a long period. The re-entrant well is filled with the ice bath water (no ice) to act as a temperature transfer medium from the triple point temperature to the device being calibrated.

The ice mantle is formed by cooling the re-entrant well until the ice builds up in the cell. This is done by filling the well with crushed dry ice or by using refrigerated probes that are inserted into the well. The probe has a tube filled with refrigerant that is cooled with an attached condensing reservoir that is itself filled with crushed dry ice and alcohol. This provides a continual flow of cold refrigerant down the tube and back to the condensing reservoir.

The detailed procedure for preparing and maintaining a triple point of water cell is described by the equipment manufacturer. The step-by-step procedure should be closely followed in order to safely produce and maintain the cell. Properly used, the cell can produce the triple point temperature with an uncertainty of +0.0°C – 0.00015°C.

The integrity of the cell can be checked by slowly rotating the cell so that the liquid fills the vapor space at the top of the cylinder. As the cell passes the horizontal position, a sharp click will be heard. This click is produced by the collapse of water vapor bubbles. The click will not occur if air is present to cushion the water as it hits the end of the cell. The click is louder if the cell is cooled. The water should almost completely compress the water vapor bubble as the cell is completely rotated. These two tests show that the cell is sealed and no air is present.

11.8 Metal Fixed Point Cells

Metal fixed point cells (see Figure 11-11) are constructed in much the same manner as the TPW; only the materials of construction are different. In the cells, a very pure (99.9999%) sample of metal is contained within a crucible that is then sealed within an argon-filled quartz cylinder. The interior pressure is set at a known pressure at the melting or freezing point of the metal and then the cylinder is sealed. Variations of the cells include open cells that can be gas purged and then maintained at a given pressure by the user. The cells are inserted into special furnaces that have good uniformity and stability characteristics in the area of the cells. There are usually preset ramp and hold temperature profiles that can be set for a particular metal cell type.

Figure 11-11. Metal Fixed Point Cells
(Courtesy, Hart Scientific)

Metal fixed point cells work on the principle that the temperature of the pure metal that occurs when the solid and liquid phases are in equilibrium is a constant at a given pressure. The freezing point cell is heated slightly above the metal's melting point until all the metal is molten. The cell is then cooled and held at a temperature slightly below the point at which the metal freezes to a solid phase. During the transition from molten to solid both phases are present and the temperature remains at a plateau. The probe being calibrated is inserted into the re-entrant well, and the output at the freeze point temperature is determined. In a well-designed furnace the plateau will be maintained for several hours, allowing multiple probes to be calibrated. The cell used for gallium is a melting point cell, and the temperature is first cooled and then maintained slightly above the melting point. The cell is also made of a flexible material to account for the large expansion that occurs as the metal cools. These metal point cells duplicate the metals used to define points on the International Temperature Scale of 1990.

Typical cells include:

- Melting point of Gallium   29.7646 °C
- Freezing point of Indium   156.5985 °C
- Freezing point of Tin   231.928 °C
- Freezing point of Zinc   419.527 °C
- Freezing point of Aluminum   660.323 °C
- Freezing point of Silver   961.78 °C

Another cell that is produced is the freezing point of copper cell, which is not a defining point on the International Temperature Scale of 1990.

- Freezing point of Copper   1084.62 °C

Another triple point cell uses Mercury as the metal, and this triple point temperature is used in ITS-90.

- Triple point of Mercury   -38.8334 °C

The uncertainties of the metal freeze and melt points range from 0.04 mK to 1.1 mK.

The cost of cells, associated furnaces, and labor are much higher for using and maintaining metal fixed point cells than they are for triple point of water cells. They are primarily used for standards laboratories that are calibrating SPRTs.

11.9 Ice Baths

An ice bath containing a properly prepared mixture of ice and water can produce a very reproducible temperature reference point of 0°C. The ice point was used as a defining point in earlier versions of the International Temperature Scale before the more accurate triple point of water was introduced. The ice point is defined as the temperature of equilibrium between ice and air-saturated water at standard atmospheric pressure.

The ice bath should be prepared in a cylindrical Dewer type flask, preferably made of glass. A wide-mouthed thermos bottle with a glass inner container offers a good alternative to the laboratory type Dewer. The flask should be approximately one-third filled with air-saturated distilled water. Ice made from distilled water is then added. The ice should be shaved or crushed to a diameter of 2-3 millimeters. Shaved ice is preferable and should have the consistency of snow. The ice is added until the flask is full, and then the mixture is compressed until it is tightly packed. The water should fill the space between the ice but its volume should not be large enough to cause temperature gradients. A siphon tube inserted into the bottom of the flask is used to remove excess water and melted ice water from the flask. All sources of potential contamination should be eliminated from the bath during its preparation and use. The Dewer and other equipment should be rinsed with distilled water before they are used, as should any probes that are inserted into the bath. Equipment, ice, and water should only be handled with rubber gloves. Probes inserted into the bath should be pre-cooled to prevent localized melting around the probe. The bath should be allowed to equalize for approximately 30 minutes before use.

A properly prepared ice bath can produce the ice point with an uncertainty of 0.002°C. A carefully prepared bath at NIST using distilled and de-ionized pure water and ice from that water produced a bath with a temperature of +0.000089°C. A similar bath made with local tap water produced a temperature of -0.0158°C. This lower temperature is the result of salts that are present in the tap water. The effect of atmospheric pressure on the ice point is relatively small and can be ignored for most applications. A change of 10 torr in atmospheric pressure will change the melting point of ice by only 0.0001°C.

11.10 Boiling Point of Water

The boiling point of water would seem to be an attractive reference point but is seldom used to accurately calibrate temperature probes. This is because of the effects of atmospheric pressure and the contamination of the water. Distilled or de-ionized water should be used since the salts present in tap water will raise the boiling point. The atmospheric pressure is affected by elevation above sea level and the hour-to-hour changes in pressure due to weather conditions. Water will boil at 100°C at sea level and 68°C at the top of Mount Everest. At Denver,

Colorado (elevation 5277 feet) water will boil at 94.7°C. A change of atmospheric pressure of 10 torr will change the boiling point by 0.37°C.

If the boiling point of distilled water is used as a reference then the atmospheric pressure in the room must be measured and the boiling point must be adjusted for this pressure. Room-to-room and inside-to-outside pressures will vary because of air conditioning or air extraction equipment. Atmospheric pressures provided by local weather stations are referenced back to sea level. This procedure removes the altitude effect, which can be greater than the weather effect.

The use of atmospheric pressure–adjusted boiling water as a calibration point should be limited to industrial temperature probes where an accuracy of ±0.1°C is required.

12

Electrical Calibration

Common electrical parameters that require calibration equipment include voltage, current, resistance, and capacitance. Equipment is needed to measure the outputs of transducers and signal conditioners and to simulate the inputs to displays, data acquisition systems, and signal conditioners.

Bench-mounted digital multimeters with resolutions ranging from five and one-half digits to eight and one-half digits will provide the accuracy necessary to calibrate most transducer-based measurement systems. These meters will typically be able to measure AC and DC voltage, AC and DC current as well as resistance, frequency, and period.

Input simulation can be performed with power supplies, current sources, resistance and capacitance boxes, and frequency or signal generators. Multi-function calibrators are also available to generate and measure electrical parameters. These calibrators and the multimeters can often be computer controlled to produce automatic calibration systems that can follow a prepared calibration procedure with and without operator input.

12.1 Voltage

DC voltage is the most common electrical parameter measured during the calibration of transducer-based measurement systems. The outputs of most signal conditioners and the output of many transducers is a DC voltage signal. The voltage can range from microvolt-level thermocouple signals to the 10-volt signals common to signal conditioners.

Five and one-half to eight and one-half digit multimeters will usually have a very accurate DC voltage specification since the DC voltage measurement forms the basis of most of the other parameters such as current, resistance, and AC voltage.

The specifications of the chosen multimeter should be carefully examined to ensure that it has the required accuracy and resolution for the particular measurement. The manufacturer may specify 24-hour, 3-month, 6-month, and 1-year accuracy specifications that will get worse as the time is increased. The 24-hour specification is an indication of the best accuracy that the multimeter is able to deliver, usually under tight ambient temperature limits. It is probably not a practical specification to use since it is only valid for the 24 hours following calibration. The 1-year specification is more realistic unless the multimeter is calibrated more frequently than once a year.

AC voltage measurements are encountered during the calibration of vibration instrumentation. Vibration transducers such as accelerometers and velocity transducers are typically calibrated by mounting the transducer and a reference transducer on an electro-dynamic shaker table that is driven by a power amplifier. The power amplifier provides a sinusoidal output that causes the shaker table and transducer to move in a proportional sinusoidal motion. The frequency of the motion is equal to the frequency of the power amplifier output. The magnitude of the motion is proportional to the power delivered to the shaker by the amplifier.

When a digital multimeter is used to measure the vibration signal, the user must remember that the multimeter indicates volts RMS (Root Mean Square). The sinusoidal motion of the transducer proportional to the RMS voltage is the RMS acceleration, velocity, or displacement.

Zero-to-peak equivalent for a pure sine wave is RMS times the square root of two, which is equal to 1.414. Peak-to-peak equivalent is twice the zero-to-peak value or RMS times 2.828.

The user should also check the specifications for the total frequency range over which the AC voltage readings are valid and the accuracies for different frequency ranges. The errors in AC voltage measurements for frequencies less than 10 hertz will often be significant when using digital voltmeters to calibrate vibration transducers.

12.2 Current

DC current is a common signal output from process instruments. All manufacturers use a standard 4-to-20-milliamp DC signal as the output proportional to the calibrated range of the process instrument. A pressure transmitter with a calibrated range of 0 to 100 psig will provide an output signal of 4 milliamps at 0 psig and 20 milliamps at 100 psig. A pressure transmitter with a range of 100 to 300 psig will provide an output signal of 4 milliamps at 100 psig and 20 milliamps at 300 psig. The use of 4 milliamps rather than 0 milliamps to represent the lowest value in the range is known as a "live zero." Using this live zero shows that the current path or current loop is complete and there are no breaks in the wiring. If a break in the wiring does occur then the current signal will drop to zero. The process operators can now tell the difference between a wiring break and a transmitter operating at the minimum value in the range.

The DC current specifications of typical five-and-one-half-digit and six-and-one-half-digit multimeters should be checked before using them as an output standard for calibrating modern process instrumentation. The typical accuracy for these instruments is 0.1% of full scale. A high-quality digital multimeter may only have a DC current accuracy of twice this value, making it useless as a calibration standard.

Higher accuracy multimeters or process instrument calibrators are a better choice for measuring the current output from transmitters. Process instrument calibrators are designed to measure the narrow 4-to-20-milliamp current signal range with high accuracy. General-purpose multimeters have to measure current over a much wider range, and their typical applications do not require this higher accuracy.

AC current is used to calibrate multimeters and clamp-on ammeters. The high-current calibration of clamp-on ammeters is carried out by passing a lower current through a multi-turn coil of wire. The jaws of the clamp-on ammeter are centered around the turns of the coil, and the ammeter responds to the total magnetic field created by the current passing through each separate wire. A 10 amp current passing through 50 turns of wire will produce the same magnetic field as 500 amps passing through one wire. The design of a calibration coil must produce a uniform magnetic field in the clamp area and must have a coil-clamp inductance that does not cause source oscillation or shut-down due to voltage compliance. It is also important to locate the coil at the position indicated by the alignment marks on the ammeter's jaws. The manufacturer of multifunction electrical calibrators will usually offer a coil designed for use with the calibrator.

12.3 Resistance

Calibrating RTDs and thermistors requires calibrators that can accurately measure the output resistance of the transducers. Digital multimeters with resolutions of five and one-half or six and one-half digits are usually accurate enough to measure the output resistance of thermistors. These sensors have a large resistance change per degree compared to RTDs, and the multimeter does not require the high accuracy and resolution that is needed for RTD calibration. RTDs have a small resistance change per degree, and even a good quality six-and-one-half-digit multimeter may not have the necessary accuracy and resolution to be used as a calibration standard for anything better than an industrial-grade RTD.

An industrial-grade 100-ohm platinum RTD will have a resistance change of approximately 40 ohms per 100°C. This results in the approximate resistance-versus-temperature table shown in Table 12-1 for a range of 0 to 300°C.

Table 12-1. Approximate Resistance-versus-Temperature for a Range of 0 to 300°C

Temperature (°C)	Resistance (ohms)
0	100
50	120
100	140
150	160
200	180
250	200
300	220

The six-and-one-half digit multimeter will likely have an over range of only 20% and on the 100-ohm range will be capable of measuring to 119.9999 ohms. This is not sufficient to reach the required 220 ohms, and so the 1000-ohm range will have to be used. This range is capable of measuring up to 1199.999 ohms with a resolution of 0.001 ohms.

The RTD output is approximately 40 ohms per 100°C.

> or 0.4 ohms per degree Celsius
>
> or 0.04 ohms per 0.1°C
>
> or 0.004 ohms per 0.01°C
>
> or 0.001 ohms per 0.0025°C

The resolution of the multimeter represents a temperature change of 0.0025°C, which is more than sufficient for an industrial-grade RTD. The question now is whether the multimeter has the required accuracy to be used as a calibration standard for the RTD.

The one-year accuracy specification for the 1000-ohm range is specified as:

$$\pm (0.01\% \text{ of reading} + 0.001\% \text{ of range})$$

For a measurement of 100 ohms on the 1000-ohm range, this is a possible error of:

$$\pm (0.01/100 \times 100 + 0.001/100 \times 1000) \text{ ohms}$$

$$= \pm (0.01 + 0.01) \text{ ohms}$$

$$= \pm 0.02 \text{ ohms}$$

This represents a temperature error of ±0.05°C, which is within the ±0.35°C error allowed for a Class A industrial RTD at 100°C.

The same type of analysis for a five-and-one-half-digit multimeter may show that the resistance accuracy is not sufficient for use as a calibration standard. The accuracy of the six-and-one-half-digit multimeter will probably not be good enough to calibrate higher-accuracy working standard RTDs. The requirement for a four-to-one TAR (Test Accuracy Ratio) between the standard and the unit being calibrated should also be taken into account.

More accurate resistance calibrations can be conducted with eight-and-one-half-digit multimeters or with resistance bridges. There are also very high accuracy RTD thermometers that are designed specifically for use with Standard Platinum Resistance Thermometers (SPRTs). These RTD thermometers provide resistance bridge accuracies over the limited resistance range of RTDs. Typical eight-and-one-half-digit multimeters will provide accuracies to 10 parts per million (ppm), while resistance bridges have accuracies equal to or better than 1 part per million. The RTD thermometers and resistance bridges have specialized automatic features like current reversal to eliminate thermal junction effects. These types of features have to be implemented manually in the case of digital multimeters.

All high-accuracy resistance measurements have to be carried out using equipment that allows four-wire inputs so as to eliminate lead-wire resistance. The meter terminals and the wire terminals should be gold plated to minimize contact resistance and thermal emfs.

The current supplied to the sensor resistive element should be approximately the same value as that supplied by the sensor signal conditioning. Too high a current will cause the sensor element to self-heat from the power generated by the current passing through the element. One-hundred-ohm RTDs typically use currents of one milliamp DC. Since Power = current^2 x resistance, the one milliamp current produces 0.1 milliwatts of power in the RTD at 100 ohms. This small amount of heat energy will be dissipated into the medium flowing past the sensor, and no resistance change due to the self-heating effect will occur. If 10 milliamps DC are used by the meter then the power increases to 10 milliwatts. This may cause a significant amount of heat energy that cannot be dissipated by the flowing medium. In this case, the resistance measurement may be in error due to the self-heating effect.

12.4 Thermocouples

Calibrating thermocouples requires that DC voltage be measured, typically in the range of -20 to +50 millivolts. Type J thermocouples have a temperature sensitivity of approximately 50 microvolts per degree

Celsius. The typical error specification for special grade J type thermocouple wire is ± 1.1°C.

The accuracy specification for a good quality six-and-one-half-digit multimeter on the 100-millivolt range is typically less than ± 5 microvolts. Since this represents a thermocouple reading of approximately ± 0.1°C we see that the meter is adequate for the calibration of this thermocouple.

Thermocouple calibrators that are specifically designed for high-accuracy DC voltage measurements are also available. These units store the curve coefficients for the individual thermocouple types and display the thermocouple outputs in temperature units as well as in millivolts. They also have built-in cold junction temperature measurement capability to account for the offset voltage caused by wire termination temperature.

12.5 Multifunction Calibrators

Multifunction electrical calibrators (see Figure 12-1) are bench- or rack-mounted units that source and/or measure most of the common electrical parameters. These units will typically duplicate the functions of six to twelve separate individual pieces of calibration equipment. They are typically small and light enough to be moved to various areas within a test laboratory or production facility. They have been designed to provide the same high-accuracy calibration signals over a wide temperature range (15 to 35°C typical) and at high humidity levels (up to 70% RH, noncondensing). These features allow the units to carry out field calibrations as well as laboratory calibrations.

A typical calibrator will be able to source the following parameters:

- DC voltage up to 1,000 volts
- DC current up to 20 amps
- AC voltage up to 1,000 volts, 10 to 500 kHz sine wave
- AC current up to 20 amps, 10 to 30 kHz sine wave
- Resistance up to 1,000 Megohms

Figure 12-1. Multifunction Electrical Calibrator
(Courtesy, Fluke Corporation)

- Capacitance up to 100 millifarads
- 10 types of thermocouples over their complete range
- 8 types of Resistance Temperature Detectors (-200 to +630°C)
- DC power to 20 kilowatts
- AC power to 20 kilowatts
- Frequency to 20 MHZ
- Phase between two simultaneous voltage and/or current outputs (0 to 180°)

The accuracy and range of these units allow them to calibrate:

- Handheld and bench multimeters up to six-and-one-half-digit resolution
- Thermocouple displays and signal conditioners
- RTD displays and signal conditioners
- Process instrument calibrators

- Recorders
- Power meters
- Clamp-on ammeters
- Portable and bench oscilloscopes
- Capacitance meters
- Data acquisition systems
- Thermocouples

Optional add-on modules allow the units to calibrate:

- Power analyzers and power quality meters
- Pressure transducers
- High-frequency oscilloscopes (up to 1 gigahertz)

The calibrators are usually set up using front-panel keypads and digitally display the value of the parameter being sourced or measured. The same display will alert the operator to any abnormal or unsafe conditions that occur. The units are also computer controlled using RS-232 or IEEE-488 interfaces. Automatic calibration procedures can be supplied or developed using calibration software. Modern test equipment such as bench multimeters can be calibrated and adjusted using the software and interfaces with little or no operator intervention.

13
Force Calibration

Force calibration equipment is needed to calibrate any of the force or load sensors described in chapter 4, section 4.7, "Force Transducers." The spans of load cells extend from a few tenths of a pound force to one million pounds force. Typical accuracies are from 0.02 to 0.1 percent of full scale. The large variation in load cell spans means that several pieces of calibration equipment would be required to cover all possible calibration requirements. The typical industrial load cell user will likely only have to calibrate a limited number of load cell ranges, those used in his or her plant. However, the test laboratory or calibration laboratory may have to purchase several calibration fixtures to cover a much wider range of load cell calibrations.

Load cells are calibrated either by applying a known force with weights or by using a hydraulic piston and a reference load standard. The load standard can be a high-accuracy load cell or a proving ring.

13.1 Dead-Weight Force Calibration

Dead-weight force calibration provides the most accurate results since it uses the fundamental unit of mass as a reference. The National Institute of Standards and Technology (NIST) uses six dead-weight calibrators to cover the range from 100 pounds force to 1,000,000 pounds force. The uncertainty is stated as 0.0005% of applied force. The overall uncertainty includes the uncertainties associated with the mass of the weights, the local acceleration due to gravity, and the determination of air density.

The air density is used to adjust the mass value of the weights for the buoyancy force that opposes the gravitational force. This buoyancy force is a direct function of the air density relative to the volume of the weight. The air density depends on the air temperature, barometric pressure, and the humidity or dew point at the time when the mass is determined.

The calibrating laboratory will state the applied force in standard pounds force. A standard pound force is the force acting on a one-pound mass under the influence of a standard gravity of 9.80665 m/s². This force is given by:

$$F = m \times g/9.80665(1\text{-air density}/\text{weight density})$$

where:

F = the standard pound force.
m = the mass of the applied weight
g = the local gravity value in units of m/sec² at the elevation of the center of gravity of the weight

Weights are typically made of stainless steel. Type 302 stainless steel has a density of 7890 kg/m³, while type 410 has a density of 7720 kg/m³. Air densities in a laboratory environment would typically be 1.1 to 1.3 kg/m³. The buoyancy correction (1-air density/weight density) for a type 302 steel weight at an air density of 1.2 kg/m³ would be 0.99985.

The force calibration machine (see Figures 13-1 and 13-2) has a rigid steel frame that supports the transducer being calibrated as well as the weights. The support allows both tension and compression calibrations to be conducted by moving the location of the transducer in the machine. The weight support and any transducer fixtures (see Figure 13-3) apply a minimum load to the transducer when there are no applied weights. This load is known as the "tare load," and it must be included in the weight load that is applied. The weights in a typical machine are added and removed by using individual hydraulic or pneumatic jacks. Other designs use a support system that sequentially picks up or drops off a series of weights as a jack raises and lowers the support. The 1,000,000-pound force machine at NIST uses twenty 50,000-pound weights to cover the calibration range.

Figure 13-1. Dead Weight Force Calibration
(Courtesy, Morehouse Instrument Company, Inc.)

A typical load cell calibration will consist of 10 equally spaced loads applied in ascending and descending order. The calibration is usually repeated, with the transducer being rotated around its applied load axis. The two calibration runs are averaged to give the final calibration results. ASTM standard E74-02 provides a recognized procedure for calibrating force-measuring instruments. The calibration is either performed as a system calibration with the associated signal conditioner and display or by calibrating the load cell separately. If calibrated separately, the output is usually measured as the ratio of the bridge output to the excitation voltage (mV/V). This technique eliminates the effects of any excitation voltage variations during calibration and allows the user to transfer the calibration to any other excitation voltage. The load cell is usually "conditioned" prior to the calibration by applying 3 load cycles from 0 to a load slightly higher than the maximum calibration load.

Load cell calibration machines (see Figure 13-4) are also available that use reference load cells instead of dead weights. The construction of

Figure 13-2. High Load Dead Weight Force Calibrator
(Courtesy, Morehouse Instrument Company, Inc.)

the machines is similar except the load is applied with a hydraulic jack and measured with a high-accuracy, calibrated load cell. The hydraulic jack is pressurized manually with a pump that is similar to a deadweight pressure tester or with a motorized pump. The accuracy of typical reference load cells is ± 0.02% of full scale. A series of reference load cells with different ranges is needed to match the ranges of the cells being calibrated.

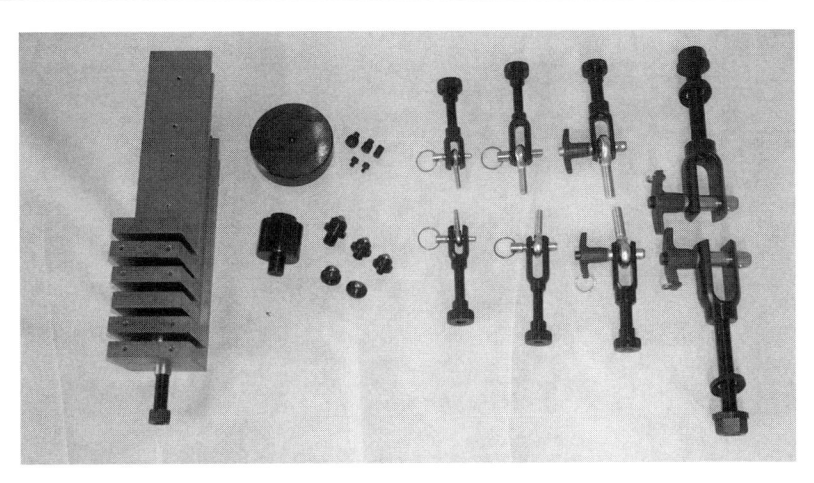

Figure 13-3. Load Cell Calibration Accessories
(Courtesy, Morehouse Instrument Company, Inc.)

These types of systems are typically used to calibrate load cells that have accuracies at or higher than 0.1% of full scale. This is usually sufficient for most industrial force measurements, but it may not be accurate enough for load cells that are used for custody transfer weighing applications.

The advantage of these systems is that they can be fully or partially automated. The electrical output of the reference load cell can be used for hydraulic jack control and can also be read by a data acquisition system along with the output from the cell being calibrated. Calibration data can be stored, and calibration results and certificates can be printed automatically.

232 Fundamentals of Test Measurement Instrumentation

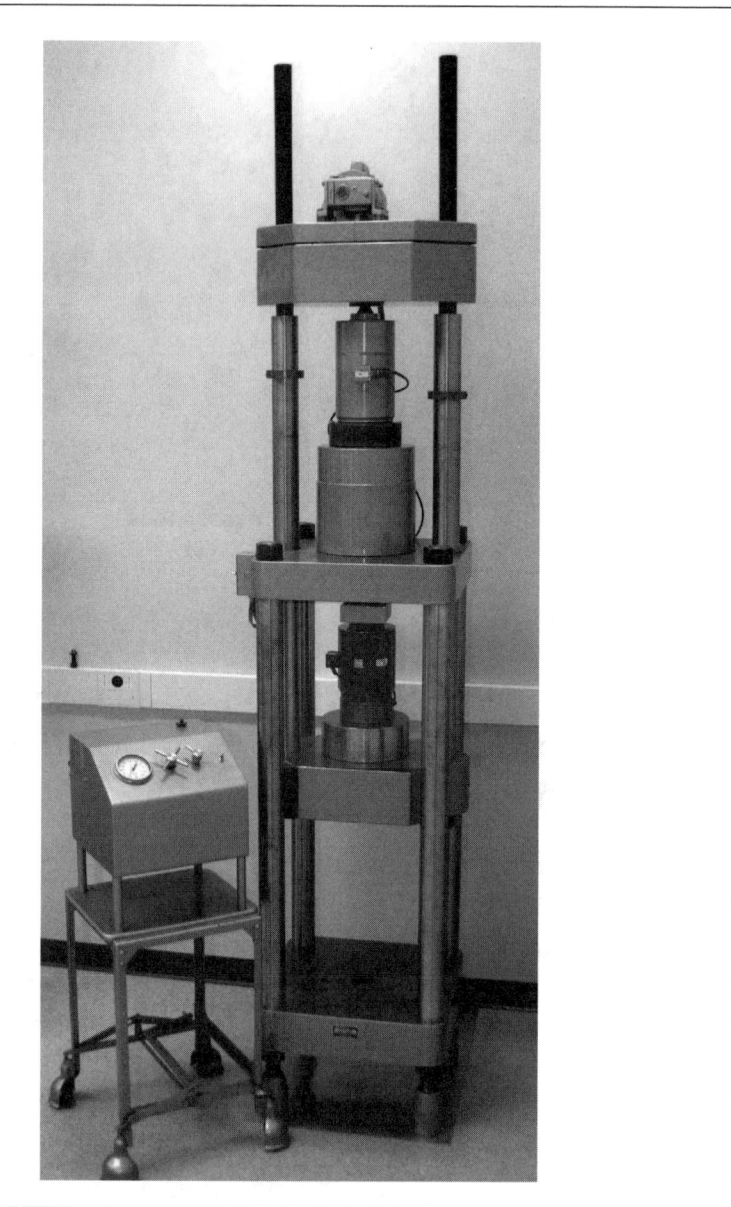

Figure 13-4. Reference Load Cell Force Calibrator
(Courtesy, Morehouse Instrument Company, Inc.)

13.2 Proving Rings

Force calibration systems similar to the reference load cell type have also been designed in which proving rings are used in place of the reference load cell. A proving ring (see Figure 13-5) consists of an elastic steel ring with an integral measuring device that measures the ring deflection as it is loaded either in tension or compression. The designed thickness, width, and diameter of the ring increases as the load capacity increases. Typical ring deflections range from 0.033 inches to 0.167 inches. Measurement uncertainties using proving rings range from 0.0125% to 0.075%. Proving rings are calibrated by using dead-weight systems (see section 13.1). The calibration basically determines the deflection of the ring as a result of applied load.

Figure 13-5. Proving Rings
(Courtesy, Morehouse Instrument Company, Inc.)

The ring deflection (change in diameter) is measured using a micrometer that is mounted inside the ring. A vibrating reed is mounted in line with the micrometer spindle and used to determine the amount of deflection. The reed is gently tapped to start it vibrating, and then the micrometer spindle is advanced until it just touches the reed. At this point, a characteristic buzzing sound will be heard to indicate contact with the reed. The micrometer dial is read at this point to determine the change in ring diameter as a result of the applied load.

Proving rings are made to operate in compression or in tension. Some models can be used in both modes. Those used to measure tension load will have threaded bosses on the top and bottom of the ring. These bosses accept accessories that mate the ring with the load cell being calibrated. These accessories include tension members to join ring and load cell with a threaded rod. Clevis and pin assemblies that join the two and also eliminate sources of misalignment are also available.

The most accurate proving rings can provide a slightly more accurate calibration than can reference load cells. The disadvantage of proving rings is that the deflection must be measured manually. This is a slower process and one that is not suitable for automation, in contrast to the reference load cell system.

13.3 Hydraulic Presses

A hydraulic press or jack (see Figure 13-6) is used in the load calibration machines to produce the force. The press or jack is essentially a vertical cylinder with a close-fitting internal piston. Oil under the piston is pressurized with a manual or motor-operated pump. The close tolerance between the piston and cylinder diameters prevents a significant amount of oil from leaking past the piston, and the piston will rise until it comes into contact with the load cell mounted in line with the jack. Once in contact, the piston can no longer move, and any additional oil pressure will cause the piston to exert a force on the load cell. This force is basically equal to the area of the piston times the applied pressure. An oil pressure of 1,000 psi acting on a piston with an area of 10 square inches will produce a force of 10,000 pounds. Typical manual pumps are similar or identical to the pumps used with working level dead-weight testers. These pumps are relatively inexpensive and have low internal leakage. They also usually have a threaded screw device that manually trims the applied pressure to the required level.

The piston extension that comes into contact with the load cell or with a mounting plate in the calibration machine is provided with a convex, hardened steel boss that eliminates axial misalignment. The boss will also have a hardened steel ring surrounding it on which steel cylinders

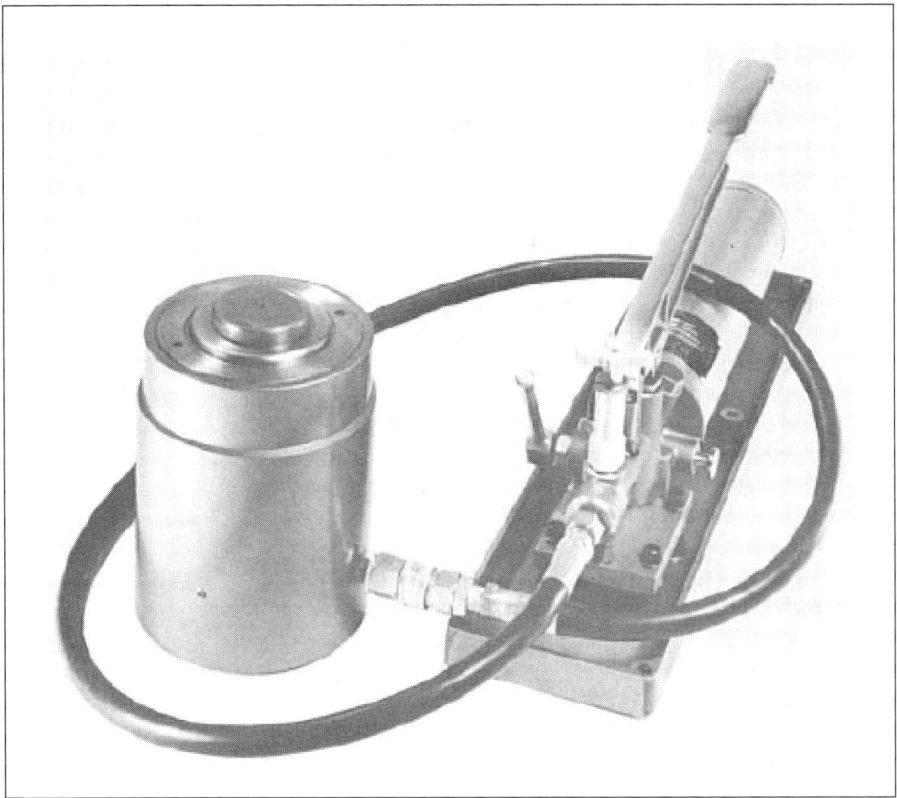

Figure 13-6. Hydraulic Press
(Courtesy, Morehouse Instrument Company, Inc.)

of various lengths can be supported. These cylinders are used to adjust the space between the piston and the various size load cells or the mounting plate.

13.4 Shunt Calibration

The shunt calibration technique was described in chapter 5, section 5.3, "Strain Gauge Amplifiers.". It applies to any strain gauge–based transducer, including load cells. The transducer is connected to the signal conditioner, and a large value resistor is connected in parallel with one of the strain gauges that makes up the Wheatstone bridge. This simulates a lowering of the resistance of that strain gauge and unbalances the bridge, causing an output change. Figure 13-7 illustrates a commercial strain gauge calibrator. The calibration is only a test of the signal conditioner and display since the transducer itself is not mechanically stressed.

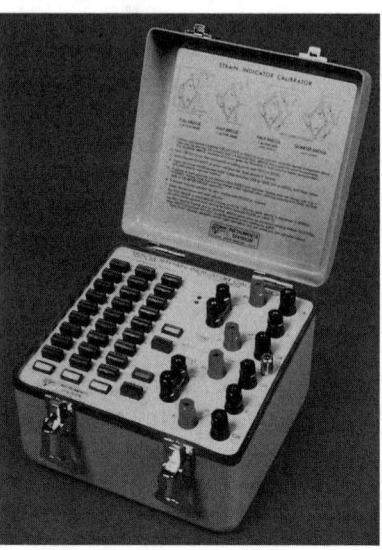

Figure 13-7. Strain Gauge Calibrator
(Courtesy, Vishay Micro-Measurements)

Ideally, the system output should be determined once the system has been calibrated and installed in the field. The resistor value is chosen to give an output of approximately 80% of the calibrated range. For subsequent checks of the system output, the load on the cell should be exactly the same as for the initial determination. If the output was first checked with no load on the cell then this should be the condition for later checks. If any pre-load was present initially then the same pre-load must be present when using the shunt calibration resistor.

Shunt calibration should be considered as a system integrity check and not as a valid calibration. The test basically gives the operator some degree of confidence that the components in the system are operating as they were during the calibration. A change in the shunt calibration output signal may indicate that zero and gain adjustments may have been changed or that the transducer may have been over ranged.

A strain gauge tester (see Figure 13-8) can be used before calibrating to verify the integrity of the strain gauge installation.

Chapter 13 – Force Calibration 237

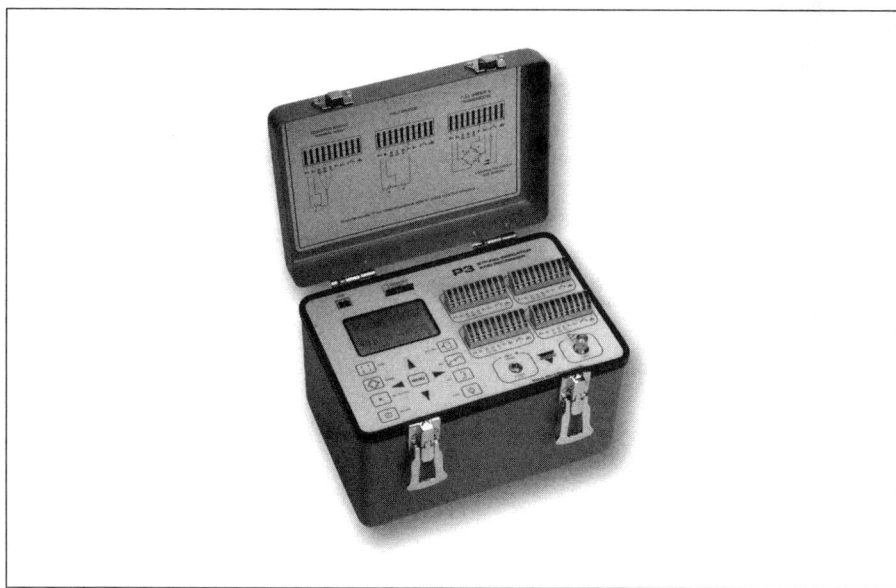

Figure 13-8. Strain Gauge Tester
(Courtesy, Vishay Micro-Measurements)

14
Flow Calibration

Calibrating flow is essential for any transducers or measurement systems that provide information on the amount of material transferred in a given time. These measurements will usually be made in units of mass per unit time (mass flow) or volume per unit time (volumetric flow). The two types of measurements are connected by the density (mass per unit volume) of the flowing material. The major factors that affect density are pressure and temperature, and these measurements are usually encountered in any flow calibration work. The majority of flowmeters in use today are mounted in circular pipes and calibration facilities have to be able to match the field installation of the devices.

The accuracies associated with flow calibration are not nearly as good as those associated with the calibration of temperature, pressure, mass, or electrical equipment. National standard and primary standard laboratories can typically calibrate temperature, pressure, mass, and electrical equipment with uncertainties ranging from 1 to 10 parts per million. In contrast, the uncertainties for NIST gas flow calibrations range from 200 to 2000 parts per million, while liquid flow calibrations are in the 800 to 1000 parts per million range.

The equipment needed for flow calibrations can be very bulky and expensive to operate. The water flow calibration facility at NIST contains an 8,000-cubic-foot water reservoir tank, 4 pumps ranging from 100 to 150 horsepower, weigh scales up to 50,000 pounds, and 30- to 40-foot piping runs with diameters ranging from 1 inch to 16 inches.

The reservoir tank alone could be 80 feet long, 10 feet deep, and 10 feet wide. A five-point calibration using this facility will cost approximately $5,000 and require a four-week turnaround time. These factors mean that most test laboratories or manufacturing plants cannot support in-house flow calibration facilities. It also means that flow equipment tends to be the least calibrated of any measurement systems.

Many flowmeters rely on having primary elements that have a specific size and/or shape to produce a signal that is proportional to flow. These include orifice plates, venturis, flow nozzles, vortex meters, and turbine meters. The physical dimensions of these elements will not change unless the flowing material induces wear or the installation procedure distorts or damages the element. An orifice plate, for example, can produce a changed differential pressure if the sharp upstream edge of the machined orifice becomes rounded due to flow-induced wear or nicked due to hard particles in the flow stream. It can also change if the installation or excessive flow causes the plate to bend. If it can be shown that dimensional values have not changed from the initial calibration then the period between actual flow type calibrations can be extended. The user should document the dimensional measurements that are made so he or she can support the claim that no changes have occurred. The balance of the flow measurement system still has to be calibrated periodically, as appropriate for the type of equipment. The differential pressure transducer that is used with an orifice plate, venturi, or flow nozzle should be calibrated at least once a year. The electronics used with a vortex meter or turbine meter can be calibrated by inputting an electrical signal with the same wave shape and frequency as the signal that is produced by the bluff body vortex sensors or turbine blades detection sensor. The dimensional measurements, differential pressure measurements, and electrical simulation measurements must be made with equipment that has traceable calibrations.

Flow calibration equipment is available for facilities that use a large number of flow transducers or are in the business of producing flow measurement equipment. These include provers for gas and liquid applications, gravimetric fluid flow calibration facilities, PVTt (Pressure, Volume, Temperature, time) gas calibrators, and reference flow calibrators.

14.1 Provers

Prover type flow calibrators are available for liquid or gas applications. The liquid prover is also known as a piston prover or positive displacement liquid flow calibrator. The gas prover is also known as a bell prover. They both use the principle of transferring a measured volume of the flowing fluid through the flowmeter in a measured time.

The piston prover for liquid calibration consists of a long stainless steel cylinder that is honed to a precise inside diameter. A close-fitting piston in the cylinder is used to displace water from the cylinder through the meter being calibrated. The piston is usually moved by air pressure on the back side of the piston. Flow rates are changed by adjusting the air pressure and using throttling valves in line with the meter being calibrated. The piston's travel is measured by a high-accuracy linear encoder that outputs a series of electrical pulses, each of which represents a very small change in piston displacement. Each pulse will be equivalent to a small change in the liquid volume that is pushed through the meter being calibrated. Knowing the time between pulses gives a measure of the volumetric flow. Pressure transducers and temperature sensors are used to measure the fluid conditions as it flows through the meter. These calibrator signals are fed, along with the output signal from the meter, to a personal computer–based data acquisition system that uses the pressure and temperature signals to calculate the density and viscosity changes of the flowing fluid. Valves are used to return the fluid from the reservoir to the cylinder after a flow run has been completed.

A prover manufacturer will usually provide several models with different cylinder diameters to cover a wide flow range. Typically, four models will allow flow rates of from 0.001 to 400 gallons per minute. Each individual model will be able to cover a 1000 to 1 maximum-to-minimum flow rate. Models are also available for flow rates of up to 3,000 gallons per minute. The calibrator fluid can be water or oil mixtures that are blended to provide fluids with different viscosities. The amount of fluid displaced during a calibration is fairly small, ranging from 1.5 gallons to 25 gallons. This small volume combined with piston stroke times of 3 to 10 seconds for maximum flow, means that the meter being calibrated must be able to make a stable measurement within the short test time. A bidirectional piston prover is also available to provide water flow as the piston moves in both directions. Diverter valves are used to maintain the flow in the same direction through the meter being calibrated. These provers are able to provide flow for longer time periods and are used to calibrate flow instruments or flow systems that cannot provide a stable output in the single stroke time.

The typical uncertainty of these piston prover calibration systems is ± 0.05% of flow.

A similar design is available to calibrate relatively low gas flow. This unit uses a series of different-diameter glass tubes with precision bores. A piston with a low friction mercury or liquid seal is used inside each tube. Gas at a constant pressure and temperature is passed through the meter being calibrated and then into the glass tube. The gas under these conditions will cause the piston to rise at a constant rate that is proportional to the gas flow. The weight of the piston provides a small, constant back pressure on the gas. The piston travel, gas pressure, and temperature are stored along with the meter output using a personal computer–based data acquisition system. The design incorporates features such as water-filled outer tube jackets to minimize variations in the bulk temperature of the gas.

The calibration range for these units is typically 1 to 24,000 standard cubic centimeters per minute. Uncertainties are approximately ± 0.25% of flow.

The bell prover, used for gas flow calibrations, consists of a vertical cylindrical inner tank that is surrounded by an outer cylindrical shell (see Figure 14-1). The annular space between the two walls is filled with a sealing fluid such as water or oil. A cylindrical "bell" (inverted tank closed at one end) is placed over the inner tank so the wall of the bell is positioned in the annular space. The liquid in the space provides a frictionless seal between the bell and the inner tank. The bell is provided with a large counterweight that balances the weight of the bell and, along with the liquid seal, allows the bell to be moved by the incoming gas under the bell with almost zero pressure drop. A second, small counterweight that moves with the bell corrects for air buoyancy effects.

Figure 14-1. Bell Prover
(Courtesy, Flow Technology, Inc.)

In operation, the calibration gas passes through the meter being calibrated and then enters the inner tank below the bell. The gas flow causes the bell to rise at a rate that is proportional to the flow. Meter output, bell displacement, gas pressure, and gas temperature measurements are stored using a data acquisition system while the bell is moving. The time required to produce a volume of gas under the displaced bell is used to calculate the gas flow. At the maximum flow of the bell prover, the bell typically moves for 5 to 6 seconds. The meter being calibrated has to be able to provide a stable output within this time period.

Several provers with different size bells are provided to suit the flow range being calibrated. These can typically range from 5 cubic foot bells to 100 cubic foot bells and can provide flow capabilities from 0.1 standard cubic feet per minute to 1,000 standard cubic feet per minute. Flow calibration laboratories have constructed bell provers that have capacities up to 10,000 standard cubic feet per minute. The typical flow rate uncertainty of bell provers is ± 0.25% of flow.

14.2 Weigh Tank

The weigh tank flow calibrator is simply described as a bucket on a scale with a stopwatch to time the filling of the bucket. In practice, the bucket becomes a large tank with a highly accurate weighing system, and the stopwatch becomes a sensor-based timing system with a crystal controlled clock. The calibrator design ensures that the flow is steady during the measurement period and that constant temperature and pressure conditions are maintained and measured. A data acquisition system controls the system and measures the test parameters. The calibration system also incorporates large reservoir tanks, pumps, diverter valves, flow straighteners, and long, straight piping runs of various diameters. The diverter valve is used to divert the normal flow path back to the reservoir onto a flow path leading into the collection tank. The flow path is reversed after a specified time or after a specified quantity of fluid has been accumulated in the weigh tank.

These types of calibrators are also known as gravimetric flow calibration systems. The term *gravimetric* refers to the fact that the flow is determined by a weight measurement. A static gravimetric system (see Figure 14-2) is one that measures how long the diverter valve is in a position that allows flow to the collection tank. It then measures the mass of the fluid in the tank after the motion of the contents has settled. A dynamic gravimetric system (see Figure 14-3) is one that measures the time it takes for the weigh tank and its contents to change from one mass value to another mass value as the flow fills the tank. Both ISO and ASME have written standards that describe the design, operation, and uncertainties of gravimetric flow calibration facilities.[2, 3]

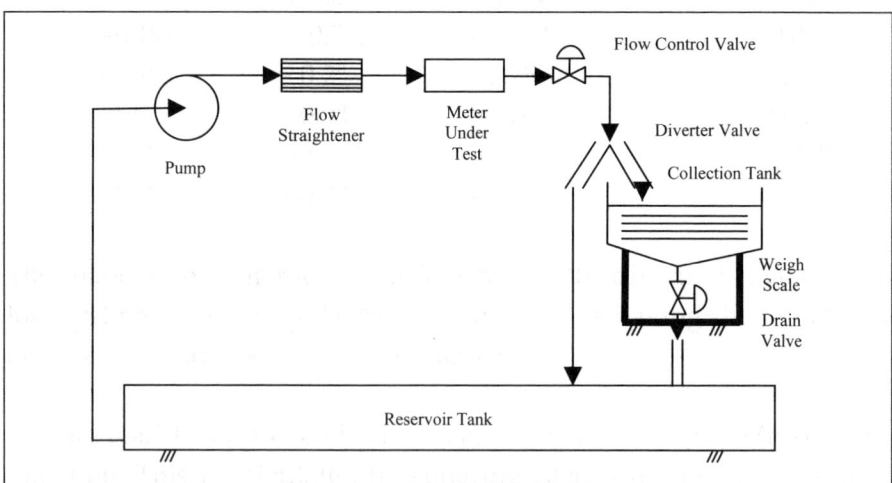

Figure 14-2. Static Gravimetric Flow Calibration Facility

---

2. *Measurement of Liquid Flow in Closed Conduits – Method by Collection of the Liquid in a Volumetric Tank*, ISO 8316: 1987 (E), International Organization for Standardization.
3. *Measurement of Liquid Flow in Closed Conduits by Weighing Method*, ASME/ANSI MFC-9M-1988, American Society of Mechanical Engineers, New York.

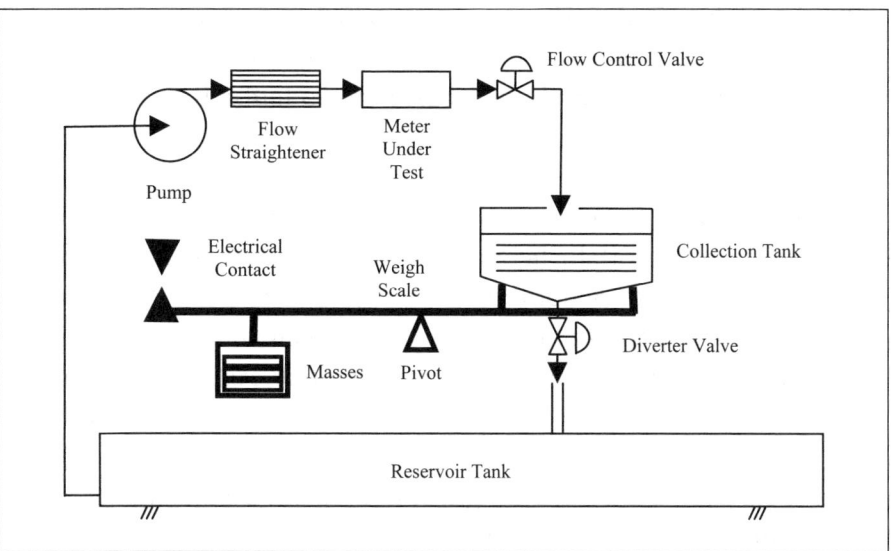

Figure 14-3. Dynamic Gravimetric Flow Calibration Facility

14.3 PVTt Calibration Facility

A Pressure-Volume-Temperature-time (PVTt) facility (see Figure 14-4) is used to calibrate gas flow meters typically in the range of 1 to 2,000 liters per minute. These calibrations are carried out to uncertainties ranging from ±0.02% of flow to ±0.05% of flow. The facility is also known as a static volumetric flow calibration system since it determines the time required to fill a tank of known volume.

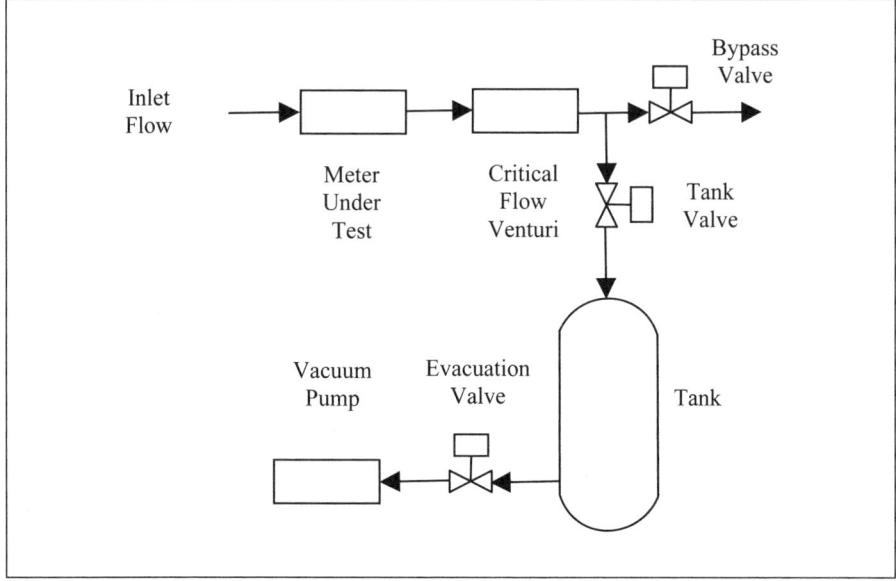

Figure 14-4. PVTt Flow Calibration Facility

The inlet gas is supplied from a compressor or from high-pressure bottled gas cylinders. The gas passes through the meter being calibrated, through a critical flow venturi, and then through the bypass valve prior to the test. The critical flow venturi is used to isolate the meter from downstream pressure variations that occur as the valves operate and the tank is evacuated or filled. The tank is evacuated with the vacuum pump, and the evacuation valve is then closed. Pressure and multiple tank temperature measurements are then taken to calculate the initial density of the tank contents. The initial mass of the tank contents can be calculated from the precisely known tank volume and the density value (mass = density x volume). The gas flow is then directed into the tank by closing the bypass valve and opening the tank valve. A timer is started, and the gas flow enters the tank until a certain pressure is achieved, at which time the valves are reversed and the timer stopped. The pressure and temperature measurements are again taken, and a final density of the tank contents is calculated. A final mass is calculated, and the difference from the initial mass becomes the mass transferred in the measured time. This mass flow rate is compared to the flow rate measured by the meter being tested.

Multiple temperature sensors are located at various places in the tank, and their outputs are averaged to determine a bulk temperature measurement of the tank's contents. Some designs use multiple, small-diameter tanks that are immersed in a circulating water bath maintained at room temperature. The gas in the tanks quickly reaches an equilibrium temperature with the surrounding water, and temperature variations within the tanks are reduced.

The pressure and temperature is also measured in the piping between the meter being tested and the tank valve. This volume of gas is known as the inventory volume, and its mass must be added to the mass of the tank contents since it has passed through the meter being calibrated.

14.4 Reference Flow Sensor

Reference flow transducers are used to calibrate working level instruments either by using the reference transducer and the working level instrument in a calibration system or by installing the reference transducer in series with the working instrument in the test rig or plant piping. The fact that both sensors require long upstream and downstream piping lengths usually makes this a difficult installation unless the piping was initially designed with the calibration function in mind.

Typical reference flow sensors include critical flow venturis, critical nozzles, positive displacement meters, and turbine flowmeters. These sensors are initially calibrated in a flow laboratory using one of the systems described in sections 14.1 to 14.3 earlier in this chapter.

Ultrasonic flow technology is progressing rapidly, to the point that it can be used as a flow calibration reference. This means that the ultrasonic transducers can be attached to the plant piping that contains the working level instrument. The newer ultrasonic flowmeters use multipath signals and a self-learning capability to better define the flow profile in real time. This allows higher accuracy volumetric flow measurements to be made using the time-of-flight information and the flow distribution information.

15
Displacement Calibration

Calibrating displacement measuring equipment usually takes place at two levels. The first level is the traditional metrology laboratory that calibrates working level devices used on a daily basis in machine shops and assembly and production facilities. These laboratories also make dimensional measurements on items produced by the production facility or on items tested in a laboratory. The second level is the use of metrology laboratory equipment to calibrate displacement transducer systems. These calibrations typically provide the static response of the sensor as a function of distance or displacement.

The basic metrology laboratory standards are artifact standards that are made in a variety of shapes and sizes with very precise dimensions. They can be blocks with two parallel faces, cylindrical plugs or pins with precise outside diameters, rings with precise inside diameters, and thread gauges to match screw or pipe threads. Other standards that actually measure dimensions include micrometers, dial indicators, calipers, and height gauges. These standards are all subject to error because of to the thermal expansion coefficient of the material from which they are made. The universal reference temperature for dimensional measurements is 68°F or 20°C. Metrology laboratories are maintained at this temperature, and any equipment or pieces to be measured are stabilized at this temperature before they are measured.

Other equipment used in metrology laboratories includes coordinate measuring machines (CMM) and laser-based measuring equipment.

The coordinate measuring machine is comprised of a solid reference base on which the part to be measured is mounted and a three-dimensional traveling frame that carries a measuring probe. This probe or stylus contacts the surface of the part and provides the three-dimensional measurements of the contact point in relation to some fixed reference point. Each additional measurement is made in relation to the same reference point. Laser-based measurement systems include triangulation sensors and interferometer systems. We described the triangulation sensor in chapter 4, section 4.4.5, "Laser Displacement Transducer." The simplest form of the laser interferometer uses a laser beam that is split into two optical paths. The one path serves as a reference beam while the other path is reflected from the item being measured. The two beams are recombined, and the phase shift caused by the differing path lengths causes an alternating light and dark interference pattern to occur. The change in light intensity from dark to light and back to dark represents a displacement change of one half the wavelength of the light source since the total light path is from the source to the object and back again. Using a Helium-Neon laser (wavelength = 633 nanometers), this represents a resolution of 316.5 nanometers or 0.0000003165 millimeters). More sophisticated interferometer systems can improve this resolution to the sub-nanometer level.

15.1 Gauge Blocks

Gauge blocks are the simplest devices available to make highly accurate calibrations of displacement transducers. These steel blocks have two opposite faces that are ground and polished to within one millionth of the nominal size. By stacking blocks almost any displacement can be achieved. A set of 81 blocks can produce over 100,000 length combinations. A typical set will have blocks ranging from 0.05 inches to 4 inches. Longer individual blocks are also available.

When using blocks to calibrate noncontact displacement transducers, the user should make sure that the target requirements of the transducer are met. These requirements may include minimum size, minimum thickness, surface finish, material, and color. The manufacturer will specify the expected target conditions and may also state errors likely when nonstandard targets are used. The user may have to manufacture a suitable target that can be added to the gauge block stack if the gauge block does not provide a suitable target. Zirconia ceramic blocks may present a better surface for some transducers (optical or laser) than do polished steel blocks.

The user should consider contact force and the effect on the block surface finish when calibrating spring-loaded displacement transducers. Excessive force when using a stack of blocks can cause errors or damage the surface of the blocks.

15.2 Dial Indicators

Dial indicators (see Figure 15-1) provide a useful method for calibrating displacement transducers when the accuracy gauge blocks provide is not required. Mechanical dial indicators are typically available that have ranges from 0.006 inches to 12 inches. The graduation on the low-range indicators is 0.00005 inches, and the graduation on the high-range indicator is 0.001 inches. The accuracy is usually ±1 graduation over the whole travel. Electronic indicators are also available that show the stem travel on a liquid crystal digital display. The range of these digital indicators is usually more limited (0.025 inches to 4 inches), and the resolution and accuracy is typically ±0.001 inches. These indicators can also have an electronic output that can be interfaced with a data acquisition system to automatically calibrate transducers.

A dial indicator can be combined with a dovetail slide to produce a manual bench-mounted calibration system. The transducer and dial indicator can easily be mounted on the base of the slide and the target can be mounted on the slider. These dovetail slides can be obtained in lengths of up to 4 feet. The screw pitch on these units is fine enough to easily make displacement settings within ±0.001 inch.

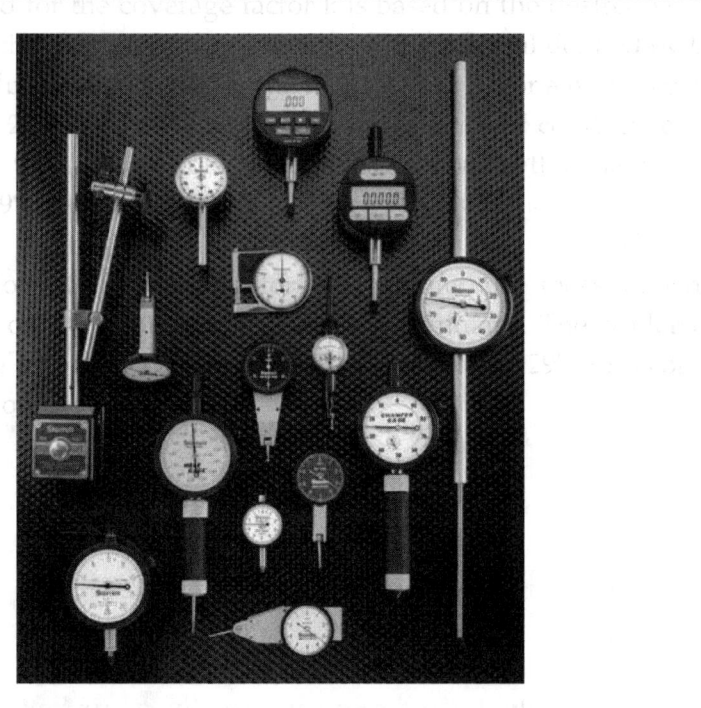

Figure 15-1. Dial Indicators
(Courtesy, The L. S. Starrett Company)

16
Vibration Calibration

Vibration is the oscillating motion of an object about an equilibrium position. The transducers that are used to measure this vibration include dynamic displacement transducers, velocity transducers, and accelerometers. The most commonly measured oscillation is sinusoidal, and calibrations are typically made with a sinusoidal source and reference. The measurement system characteristics that are determined during a calibration include low- and high-frequency limits, amplitude versus frequency, phase shifts, and identification of resonances.

Figure 16-1 shows the results of a calibration of an AC responding or dynamic transducer such as a piezoelectric accelerometer. The transducer output is compared to the output of the reference standard, which would have a perfectly flat output amplitude over the frequency range being calibrated (10 to 580 Hz). The amplitude output of the transducer is divided by the output amplitude of the reference standard at each frequency, and the result is expressed in decibels.

Result (decibels) = 20 $\log_{10}$ (transducer output/reference output)

For the graph in Figure 16-1, the output units and the calibration slope of the two instruments are the same. When the output of the transducer is the same as the reference standard, the ratio of the outputs is equal to one. Since $\log_{10}(1) = 0$, then the result in decibels is 0. In the graph in Figure 16-1 this occurs for frequencies from 60 Hz to 450 Hz. For frequencies less than 60 Hz, the transducer's output amplitude is

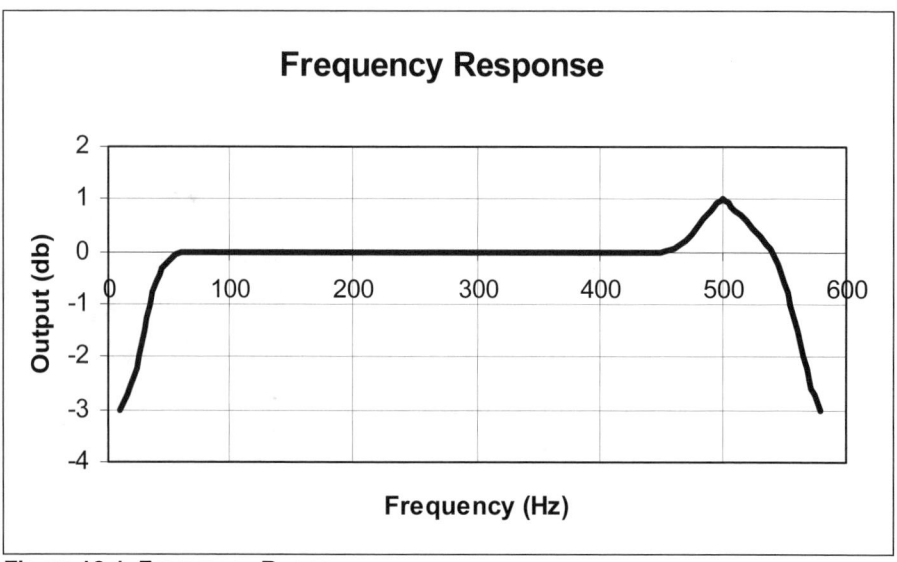

Figure 16-1. Frequency Response

less than that of the reference standard. This results in the decibel values being negative, as shown for frequencies from 10 to almost 60 Hz.

At 10 Hz, the amplitude of the transducer is 0.707 times that of the reference transducer. This results in a decibel value of -3 db. At 500 Hz a resonant peak of 1 db appears. This occurs because the testing frequency matches the resonant frequency of the transducer system. This causes the transducer's output to be higher than the output of the reference standard. The transducer output would have to be 1.122 times higher than the standard output to create this +1 db peak (20 $\log_{10}$ 1.122 = +1 db).

Above 500 Hz, the transducer output decreases until 580 Hz, where the decibel value is -3 db. Again this -3 db means that the output of the transducer is only 0.707 times that of the reference standard (20 $\log_{10}$ 0.707 = -3 db).

The specifications for this transducer would probably state that the response was flat from 10 to 580 Hz and may or may not identify the resonant peak at 500 Hz. It is typical to define the response as flat between the frequencies where the amplitude ratio is -3 db. The user should be aware that this -3 db represents an amplitude error of almost 30% (0.707). The frequencies from 60 Hz to 450 Hz represent the range where no error occurs (amplitude ratio = 0 db). In practice, a low-pass filter would be used in the signal conditioning to limit the system output's frequency to less than 450 Hz. This would eliminate the error caused by the resonant peak. A high-pass filter may be used to limit the low frequency to some value above 10 Hz. The amplitude roll-off of the filter would be known, and the system output could be defined by the filter rather than by the transducer.

The phase response of the transducer system can also be compared to that of the reference standard. The reference standard will have no phase shift compared to the actual physical vibration over the frequency range being calibrated. The phase shift of the transducer system is measured at each test frequency and expressed as the number of degrees that the system lags or leads the reference system. Signal conditioning filters and amplifier stages will usually cause some phase shifts. The mechanical response of the sensor element will also cause phase shifts at some low- and high-frequency values.

Plant personnel conducting vibration testing on components must have an understanding of phase shifts. In this case, the user may want to know how one part of the equipment is moving in time relation to another part. Using two transducer systems with different phase responses will introduce time errors into this analysis. The frequency analysis should be limited to the range where there is zero phase shift between the two transducers. In many cases, this range is smaller than the frequency range where the amplitude response is 0 db.

The amplitude of sinusoidal motion or vibration can be expressed as peak-to-peak, peak (0 to peak), rms (root mean square), or avg (average). The relationship between these values as shown in Figure 16-2 is:

$$\text{peak} = 1$$

$$\text{peak to peak} = 2 \times \text{peak}$$

$$\text{rms} = 0.707 \times \text{peak}$$

$$\text{avg} = 0.637 \times \text{peak}$$

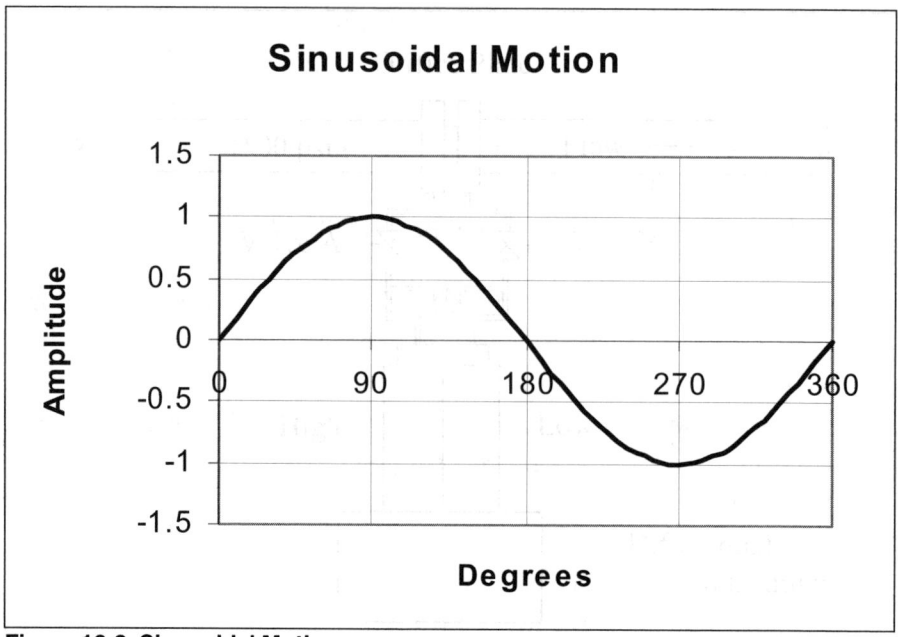

Figure 16-2. Sinusoidal Motion

Table 16-1 shows how to convert from one amplitude value to any of the other values.

Table 16-1. Converting One Amplitude Value to Other Amplitude Values

Multiply → To Obtain ↓	Peak by	Peak-to-Peak by	Average by	RMS by
Peak	1	0.5	1.57	1.414
Peak-to-Peak	2	1	3.14	2.828
Average	0.637	0.319	1	0.901
RMS	0.707	0.354	1.11	1

An object's vibration can be expressed as its acceleration, velocity, or displacement. Acceleration is the derivative of velocity with respect to time, while velocity is the derivative of displacement with respect to time. Conversely, displacement is the integral of velocity, and velocity is the integral of acceleration.

For sinusoidal vibration, the three terms are related as follows:

Displacement = velocity / (2 × Pi × frequency)
= acceleration / (2 × Pi × frequency) (2 × Pi × frequency)

Velocity = acceleration / (2 × Pi × frequency)

Acceleration = velocity × (2 × Pi × frequency)
= displacement × (2 × Pi × frequency) × (2 × Pi × frequency)

Velocity = displacement × (2 × Pi × frequency)

These relationships assume that the units are consistent and the frequency is expressed in Hz.

That is:

Displacement is in units of inches peak to peak.

Velocity is in units of inches per second peak to peak.

Acceleration is in units of inches per second squared peak to peak.

Or:

Displacement is in units of meters peak to peak.

Velocity is in units of meters per second peak to peak.

Acceleration is in units of meters per second squared peak to peak.

Table 16-2 shows the relationships between acceleration, velocity, and displacement for a range of frequencies and a number of specific acceleration values. The table illustrates how difficult it is to calibrate accelerometers at the low frequencies encountered in mechanical vibration testing. To generate 1 g acceleration at a frequency of 1 Hz requires a displacement of almost 10 inches. The same 10 inches of displacement is required to generate 100 g acceleration at 10 Hz.

Table 16-2. Relationships between Acceleration, Velocity, and Displacement

Frequently		Acceleration		Velocity	Displacement
(Hz)	(radians/sec)	(in/sec^2) pk-pk	(g) pk-pk	(in/sec) pk-pk	(in) pk-pk
1	6.28	386.09	1	61.45	9.780
5	31.42	386.09	1	12.29	0.391
10	62.83	386.09	1	6.14	0.098
20	125.66	386.09	1	3.07	0.024
30	188.50	386.09	1	2.05	0.011
40	251.33	386.09	1	1.54	0.006
50	314.16	386.09	1	1.23	0.004
60	376.99	386.09	1	1.02	0.003
1	6.28	3860.9	10	614.48	97.798
5	31.42	3860.9	10	122.90	3.912
10	62.83	3860.9	10	61.45	0.978
20	125.66	3860.9	10	30.72	0.244
30	188.50	3860.9	10	20.48	0.109
40	251.33	3860.9	10	15.36	0.061
50	314.16	3860.9	10	12.29	0.039
60	376.99	3860.9	10	10.24	0.027
1	6.28	38609	100	6144.81	977.977
5	31.42	38609	100	1228.96	39.119
10	62.83	38609	100	614.48	9.780
20	125.66	38609	100	307.24	2.445
30	188.50	38609	100	204.83	1.087
40	251.33	38609	100	153.62	0.611
50	314.16	38609	100	122.90	0.391
60	376.99	38609	100	102.41	0.272

Table 16-3 shows the accelerations that are produced for three specific displacement values and a range of frequencies.

Table 16-3. Accelerations Produced for Three Specific Displacement Values

Frequency		Displacement	Velocity	Acceleration	
(Hz)	(radians/sec)	(in) pk-pk	(in/sec) pk-pk	(in/sec^2) pk-pk	(g) pk-pk
1	6.28	0.010	0.06	0.39	0.001
5	31.42	0.010	0.31	9.87	0.026
10	62.83	0.010	0.63	39.48	0.102
20	125.66	0.010	1.26	157.91	0.409
30	188.50	0.010	1.88	355.31	0.920
40	251.33	0.010	2.51	631.65	1.636
50	314.16	0.010	3.14	986.96	2.556
60	376.99	0.010	3.77	1421.22	3.681
1	6.28	0.100	0.63	3.9	0.010
5	31.42	0.100	3.14	98.7	0.256
10	62.83	0.100	6.28	394.8	1.023
20	125.66	0.100	12.57	1579.1	4.090
30	188.50	0.100	18.85	3553.1	9.203
40	251.33	0.100	25.13	6316.5	16.360
50	314.16	0.100	31.42	9869.6	25.563
60	376.99	0.100	37.70	14212.2	36.811
1	6.28	1.000	6.28	39	0.102
5	31.42	1.000	31.42	987	2.556
10	62.83	1.000	62.83	3948	10.225
20	125.66	1.000	125.66	15791	40.901
30	188.50	1.000	188.50	35531	92.027
40	251.33	1.000	251.33	63165	163.603
50	314.16	1.000	314.16	98696	255.630
60	376.99	1.000	376.99	142122	368.107

The nomograph in Figure 16-3 illustrates graphically the relationship between acceleration, velocity, and displacement for sinusoidal vibration. Note that the units for acceleration and velocity are given in peak values, while the units for displacement are given in peak-to-peak values.

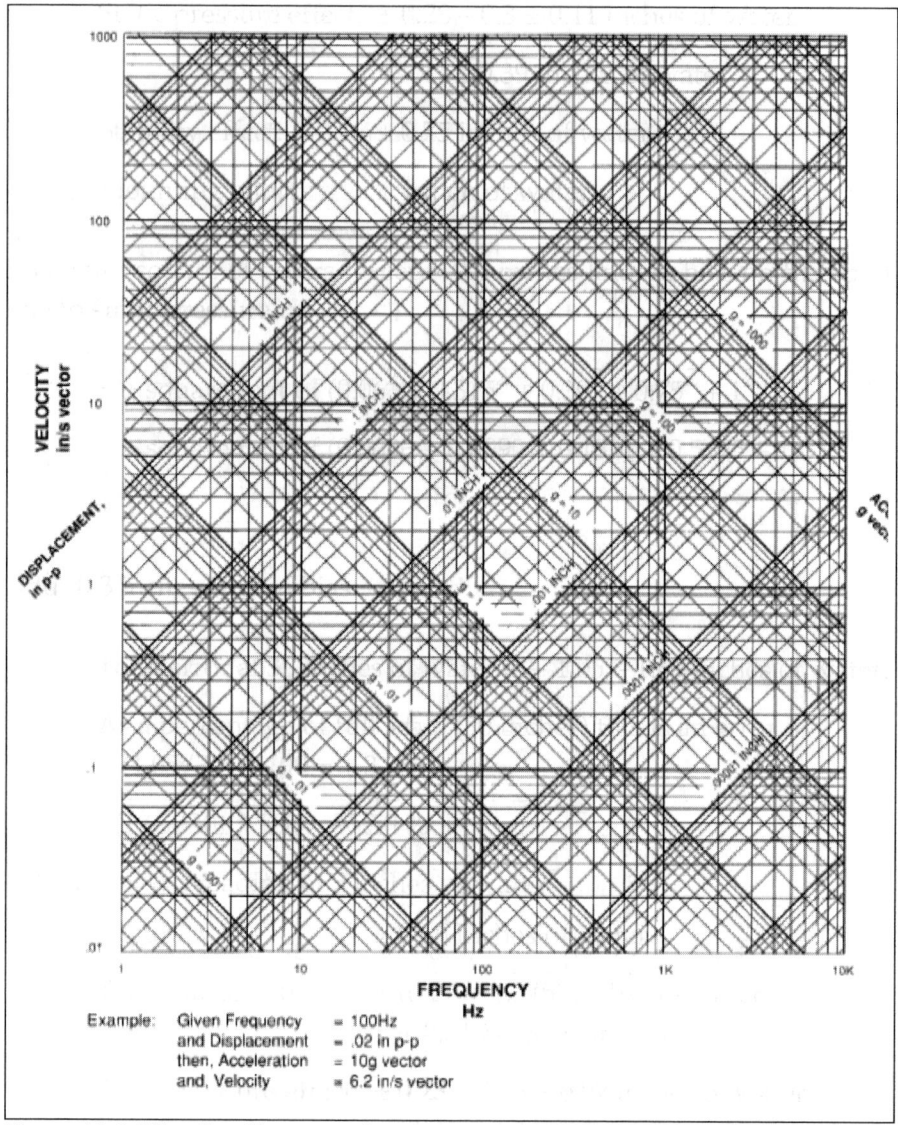

Figure 16-3. Vibration Nomograph

16.1 Shaker Table

An electro-dynamic shaker table (see Figure 16-4) is usually used to calibrate vibration transducers. These tables are typically capable of displacements of up to one half an inch, velocities of up to 50 inches per second, and accelerations of up to 50 g over a frequency range of 2 to 20,000 Hz. The coil of the shaker is driven by a variable frequency power amplifier (see Figure 16-5), which is controlled by an internal or external frequency source. A reference transducer is mounted on the table or is an integral part of the shaker's armature. The system can be operated manually or can be made as an automatic system with frequency sweeps, vibration level control, and data acquisition of the outputs from the test transducer and reference transducer. Each shaker has a maximum allowable load that can be mounted on the table. This will include the transducer, mounting fixtures, and possibly the reference transducer, if it is not built in. The maximum load is also limited by the operating frequency at the higher frequencies. The shaker and often the power amplifier are air-cooled to remove heat generated by the coil and the amplifier's electronic components.

Figure 16-4. Shaker Tables
(Courtesy, LDS Test and Measurement)

Figure 16-5. Shaker Table Power Amplifier
(Courtesy, LDS Test and Measurement)

Shaker tables are available that have larger payloads, larger operating ranges, and larger operating frequencies. The power amplifiers for these shakers become proportionally larger, as do the cooling requirements.

Beside the basic operating ranges, the important characteristics are harmonic distortion and resonant frequencies. The harmonic distortion defines how well the shaker is able to reproduce a pure sine wave motion over the whole range of the operating frequency. The first resonant frequency of the shaker system should be well above the maximum operating frequency.

Electro-hydraulic shaker tables may be required for displacements above a few inches. These systems use an oil- or water-driven piston to move the table. A servo-controlled pressure/flow hydraulic system is used to control the stroke of the piston.

Noncontact displacement transducers can also be calibrated using a shaker table. A suitable target is mounted on the table, and the transducer is mounted above the target. The mounting bracket should be stiff enough to support the transducer and should be resonance free in the frequency operating range.

16.2 Reference Accelerometer

You should use a reference accelerometer and accompanying signal conditioning amplifier as a calibration standard when using shake tables to calibrate vibration transducers. The accelerometer and amplifier are usually calibrated as a system using a laser interferometer as a reference. The amplifier typically has built-in integrators to provide velocity and displacement outputs as well as the acceleration output.

Some shaker table systems have the reference accelerometer built into the table. In other systems, the reference accelerometer and the transducer being calibrated are mounted back to back onto the shaker table. The reference accelerometer usually has internal threads at both ends of the body so that studs can join the accelerometer to the table and also to the transducer being tested.

17

Strain Gauge Calibration

Strain gauge–based transducers should be calibrated by applying the appropriate load to the transducer and measuring the resulting output. Strain gauge pressure transducers should be calibrated by applying pressure. Strain gauge load cells should be calibrated by applying dead-weight loads or hydraulic piston loads monitored by a proving ring or reference load cell. Strain gauge torque transducers should be calibrated by applying dead-weight loads to a torque arm or by using a torque wrench calibration machine.

When strain gauges are attached to equipment to determine stress levels and distributions, the ideal calibration method is to load the equipment at a known stress level and stress distribution and to monitor the output of the installed strain gauges. In many cases, this method is not practicable, and the user must rely on the shunt calibration method. Shunt calibration is carried out by placing a large value resistor in parallel with one of the arms of the strain gauge's Wheatstone bridge to simulate a reduction in the resistance of that arm and a resulting unbalancing of the bridge. This is a simulation of the applied load, and the resulting strain gauge output is calculated from the known characteristics of the gauge and the measurement system. Since this method is a simulation, it can never be considered a true calibration since there is no standard used as a reference. The method is suitable for the majority of strain gauge applications where the user wants to demonstrate that the operation of the equipment produces stresses that are well below the level where material distortion or fracture could occur.

Knowing the absolute value of the stress to a very high degree of accuracy is seldom required.

The shunt calibration method is often used to calibrate the signal conditioning and display portions of a strain gauge–based measurement system. The calibration in this case can be carried out by using the installed strain gauge bridge, by using a bridge that has known characteristics, or by using a commercial strain gauge system calibrator that has an accurate bridge, a calibrated bridge excitation voltage, and calibrated shunt resistors. This can be considered a true calibration of the signal conditioning and display since it is based on the known electrical values of the resistors and supply voltage and not on the mechanical stress applied to the equipment.

17.1 Bridge Shunt Calibration

While calibrating bridge shunts is simple in concept, the user must consider the following potential sources of error:

- Which bridge arm is to be shunted to produce a desired output?
- Where is the shunt resistor to be placed in the bridge circuit?
- What value of shunt resistor will produce the desired output?
- What is the effect of bridge nonlinearity when simulating high strain levels?

The user should also make the following assumptions:

- The excitation voltage is supplied by a constant voltage power supply (equations for constant current supply are different when nonlinearity is present).
- The input impedance of the instrument attached to the bridge output is essentially infinite.
- There are no auxiliary resistors (temperature compensation or span adjustment resistors) in the bridge circuit or the supply voltage circuit.

The basic shunt calibration techniques apply to the simple Wheatstone bridge circuit shown in Figure 17-1. In this case, no considerations need to be paid to lead wire resistance or bridge nonlinearity.

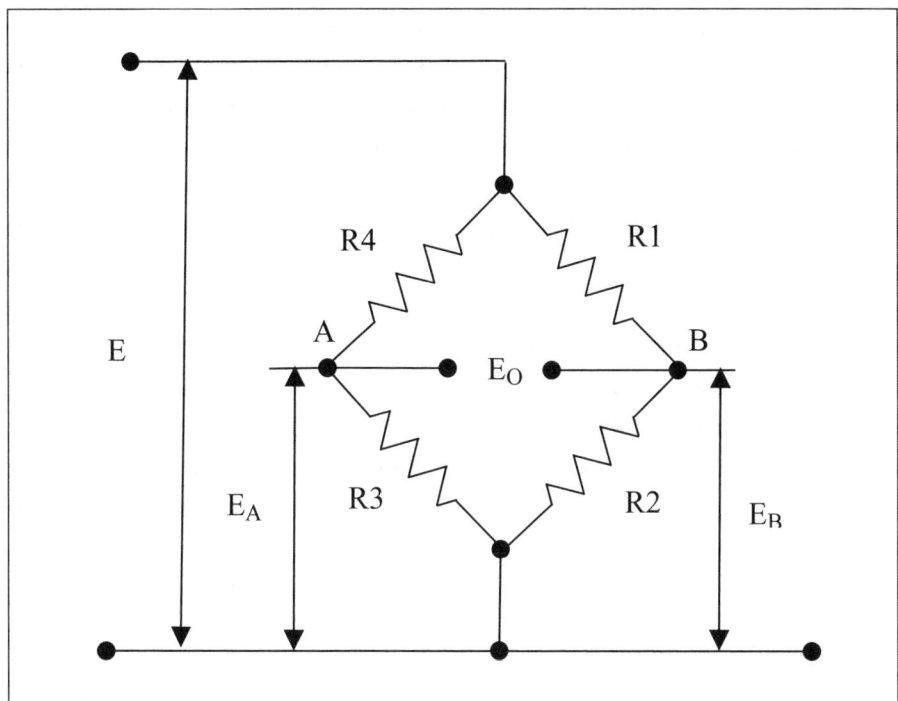

Figure 17-1. Basic Wheatstone Bridge

E – Excitation Voltage

E_O – Output Voltage = voltage difference between point A (E_A) and point B (E_B)

$E_A = E [1 - R4 / (R4+R3)]$

$E_B = E [1 - R1 / (R1+R2)]$

$E_O = E_A - E_B = E \{[1 - R4 / (R4+R3)] - [1 - R1 / (R1+R2)]\}$

$\quad = E \{[1 - R4 / (R4+R3)] - 1 + R1 / (R1+R2)]\}$

$\quad = E [R1 / (R1+R2) - R4 / (R4+R3)]$

In a nondimensional format:

$E_O / E = (R1 / R2) / [(R1 / R2) + 1] - (R4 / R3) / [(R4 / R3) + 1]$

This form of the equation shows that for a given excitation voltage the bridge output depends on the resistance ratios R1/ R2 and R4/ R3. When R1/ R2 = R4/ R3, the output is zero and the bridge is resistively balanced. The equation allows us to calculate the output voltage that will occur when any of the arms are resistively reduced by shunting. A negative change in output will occur when R1/ R2 is reduced by shunting R1 or when R4/ R3 is increased by shunting R3. A positive change will occur when R1/ R2 is increased by shunting R2 or when R4/ R3 is decreased by shunting R4.

The most common bridge configuration for strain measurements is the equal arm bridge in which R1 = R2 = R3 = R4. This circuit could have one active gauge (R1), which is shunted with a calibration resistor R_C. The bridge is assumed to initially be balanced and the lead wires short enough to produce no error.

The resistance of the arm when R1 is shunted with R_C becomes:

$$(R1 \times R_C) / (R1 + R_C)$$

The change in arm resistance is:

$$\text{Delta R} = [(R1 \times R_C) / (R1 + R_C)] - R1$$

$$\text{Or Delta R} / R1 = - R1 / (R1 + R_C)$$

The unit change in the gauge is related to strain through the definition of the gauge factor F_G. The gauge factor is given by the gauge manufacturer. A typical value is 2.000.

Delta R / R_G = F_G × strain, where R_G is the gauge resistance (typically 120 or 350 ohms).

Combining the last two equations and replacing R1 with R_G we obtain:

$$F_G \times \text{strain} = - R_G / (R_G + R_C)$$

Or:

Compressive strain obtained by shunting R_G with R_C is:

$$\text{Strain} = - R_G / F_G (R_G + R_C)$$

And:

$R_C = R_G / (F_G \times \text{strain}) - R_G$ (the first negative sign is omitted since the simulated strain is negative)

When the simulated strain is in the usual units of microstrains (strain × 10^{-6}):

$$R_C = R_G \times 10^6 / (F_G \times \text{microstrains}) - R_G$$

As stated, the analysis above is valid for small strains, constant gauge factor, and negligible resistance lead wires. The reader is directed to the reference at the bottom of this page for a more exhaustive treatment of shunt calibration.[4]

Table 17-1 shows the shunt resistors that are required to produce specific simulated values of microstrains for the general case just described. The gauge factor is assumed to be 2.000.

Table 17-1. Shunt Resistors Required to Produce Specific Simulated Values of Microstrains

Gauge Resistance (R_G) (ohms)	Shunt Resistance (R_C) (ohms)	Simulated Strain (microstrains)
120	599,880	100
120	119,880	500
120	59,880	1000
120	29,880	2000
120	19,880	3000
120	14,880	4000
120	11,880	5000
120	5,880	10000
350	349,650	500
350	174,650	1000
350	87,150	2000
350	57,983	3000
350	43,400	4000
350	34,650	5000
350	17,150	10000

---

4. http://www.vishay.com/company/brands/measurements-group/guide/tn/tn514/514intro.htm

18
System Accuracy

The overall accuracy of a measurement system depends on the accuracy of the system's individual components, the accuracy of the standards used in calibration, the factors that affect the accuracy of the components and standards, and even the actions of the person making the measurement.

We discussed the accuracy of the individual components in chapter 2, section 2.6, "Accuracy," and the factors that affect the accuracy in section 2.8, "Environmental Effects on Transducer Characteristics." A detailed analysis of how to deal with these factors is given in chapter 19. The accuracy of components or systems is always judged in relation to some known reference or standard. This standard has some stated accuracy that must be included in the overall accuracy of the component or system. In some measurement situations, the equipment operator can influence the value of the measurement. This can happen when a person views a dial indicator from an angle that causes an error in the reading (parallax error). It can also happen when a person views a rapidly changing digital display and mentally averages the readings to arrive at a single measurement. Another person making the same observations may come to a different result. These actions are called operator errors, and they should be included in the overall determination of accuracy if they can be estimated or quantified.

The methods used to combine several different sources of error or accuracy into a single accuracy figure are based on some applications of statistics. These applications have been formalized in the past few years into a determination of measurement uncertainty, which is described in written international and national standards[5] and has become a required process for calibration laboratories. These laboratories state that their calibrations are traceable to national or international standards and that the results are accurate within a calculated measurement uncertainty. The bodies that accredit these calibration laboratories will examine their measurement uncertainty analysis and use the analysis to state the laboratory's ability to provide calibrations within a certain accuracy. The measurement uncertainty of the calibrated standard is included in the measurement uncertainty of the calibrations conducted by the owner of the standard. This provides the unbroken chain of calibration that enables the test measurement made by a calibrated transducer to be traced back to the national or international standards.

18.1 System Accuracy Statistics

When a transducer manufacturer states an accuracy value it does not mean that every transducer of this model has exactly that stated accuracy. Most of the transducers will have a better accuracy than the stated accuracy, and a few will have a worse accuracy. The manufacturer will have calibrated a sample of the transducers when they were first manufactured and then calculated an expected accuracy for the rest of the transducers. Once this is done the manufacturer may not calibrate every transducer made but will instead rely on calibrating a small sample as production continues to ensure that the quality control system is maintaining the stated accuracy.

The manufacturer determines from the initial sample that the calibration results follow a normal distribution. This distribution is characterized by the familiar bell-shaped curve that shows that most of the results tend to fall in the center portion of the curve, and a few results fall at either extreme end of the curve. The form of the curve shows the manufacturer that the results of the sample calibration will be the same as the results of calibrating every transducer. The manufacturer can

---

5. ISO, "Guide to the Expression of Uncertainty in Measurement," (International Organization for Standardization, Geneva, Switzerland, 1993). B. N. Taylor and C. E. Kuyatt, "Guidelines for Evaluating and Expressing the Uncertainty of NIST Measurement Results," NIST TN-1297, 1994.

then state with confidence that the large majority of the transducers will meet a stated accuracy value and only a very few will not meet this accuracy figure. The manufacturer then carries out a risk assessment to weigh the costs associated with selling a few transducers with poorer accuracy against the cost of having to calibrate every transducer and scrapping or reworking those transducers. The transducer user can have confidence that the purchased transducer will likely meet the accuracy value but should be aware that there is a small possibility that the accuracy may be worse. The user can only determine the specific transducer's accuracy by calibrating it. The user should also be aware that there is a good possibility that the purchased transducer will have a better than stated accuracy, and this fact can be used to improve the measurement accuracy. Again, this can only be determined and quantified by calibrating.

Table 18-1 shows the accuracy values of a sample of sixty four randomly selected transducers that have been calibrated by the manufacturer. In reality, there is some variation in these results. The transducers that are shown as having an accuracy of 0% could really have accuracies from -0.025% to +0.025%. These accuracy values in the table have been grouped together and assigned the center value of 0%.

Table 18-1. Accuracy Values of 64 Transducers

Transducer Number	Accuracy % FS	Transducer Number	Accuracy % FS	Transducer Number	Accuracy % FS	Transducer Number	Accuracy % FS
1	0	17	0.05	33	0.2	49	-0.1
2	0	18	0.05	34	0.2	50	-0.1
3	0	19	0.05	35	0.22	51	-0.1
4	0	20	0.1	36	0.22	52	-0.1
5	0	21	0.1	37	0.27	53	-0.1
6	0	22	0.1	38	-0.27	54	-0.1
7	0	23	0.1	39	-0.22	55	-0.1
8	0	24	0.1	40	-0.22	56	-0.05
9	0	25	0.1	41	-0.2	57	-0.05
10	0	26	0.1	42	-0.2	58	-0.05
11	0.05	27	0.15	43	-0.2	59	-0.05
12	0.05	28	0.15	44	-0.15	60	-0.05
13	0.05	29	0.15	45	-0.15	61	-0.05
14	0.05	30	0.15	46	-0.15	62	-0.05
15	0.05	31	0.15	47	-0.15	63	-0.05
16	0.05	32	0.2	48	-0.15	64	-0.05

Table 18-2 shows a compilation of the accuracy results, showing how many transducers have a certain accuracy value (within a certain variation).

Table 18-2. Distribution of Transducer Accuracy Results

Accuracy % FS	Number of Units
0.2700	1
0.2200	2
0.2000	3
0.1500	5
0.1000	7
0.0500	9
0.0000	10
-0.0500	9
-0.1000	7
-0.1500	5
-0.2000	3
-0.2200	2
-0.2700	1

Figure 18-1 shows a plot of the results shown in Table 18-2. This is the distribution of the accuracy values and shows a normal distribution or bell-shaped curve.

The normal distribution allows us to calculate certain statistical functions that we can then apply to the expected results of all the transducers. The first of these functions is the variance, which is an indication of the spread of the results around the center value. Figures 18-2, 18-3, and 18-4 show the possible results of plotting the accuracy distributions. The plots are for the same number of transducers, and the areas under the curves are equal. Figure 18-2 shows a plot that has a shape similar to our data, which produced the plot shown in Figure 18-1. Figure 18-3 shows a distribution with a smaller spread and hence a smaller variance. Figure 18-4 shows a wider spread in the results and hence a larger variance.

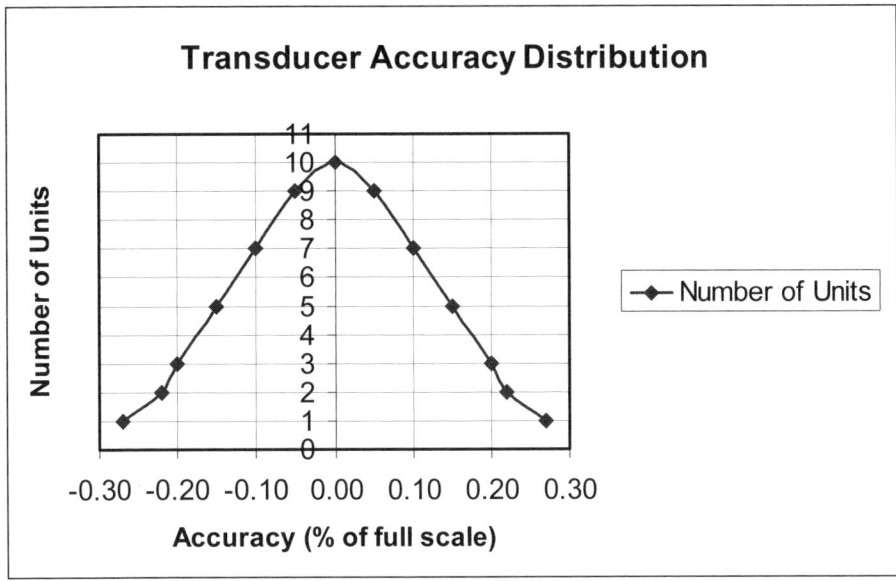
Figure 18-1. Transducer Accuracy Distribution

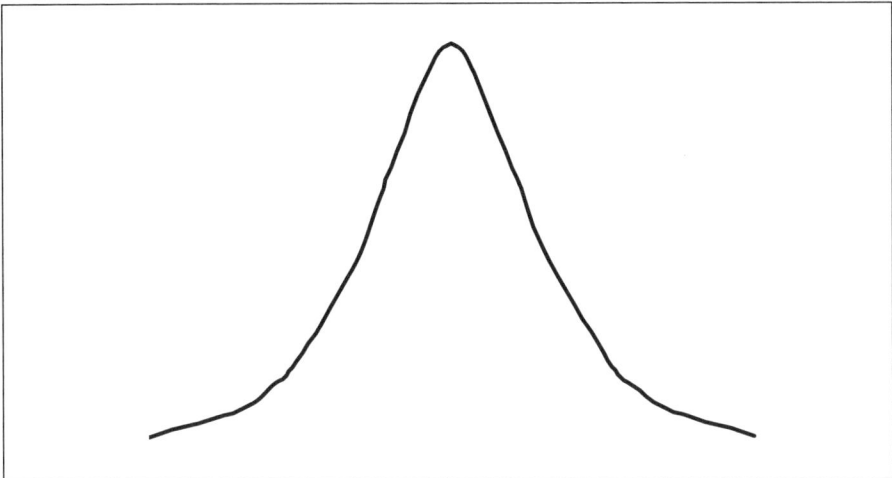
Figure 18-2. Medium Variance Accuracy Distribution

The sample variance is a number that is calculated from the data shown in Table 18-1.

$$\text{Variance} = \Sigma(x-\text{mean})^2 / (N-1)$$

where:

    x is each accuracy value
    mean is the average of the values
    N is the number of values

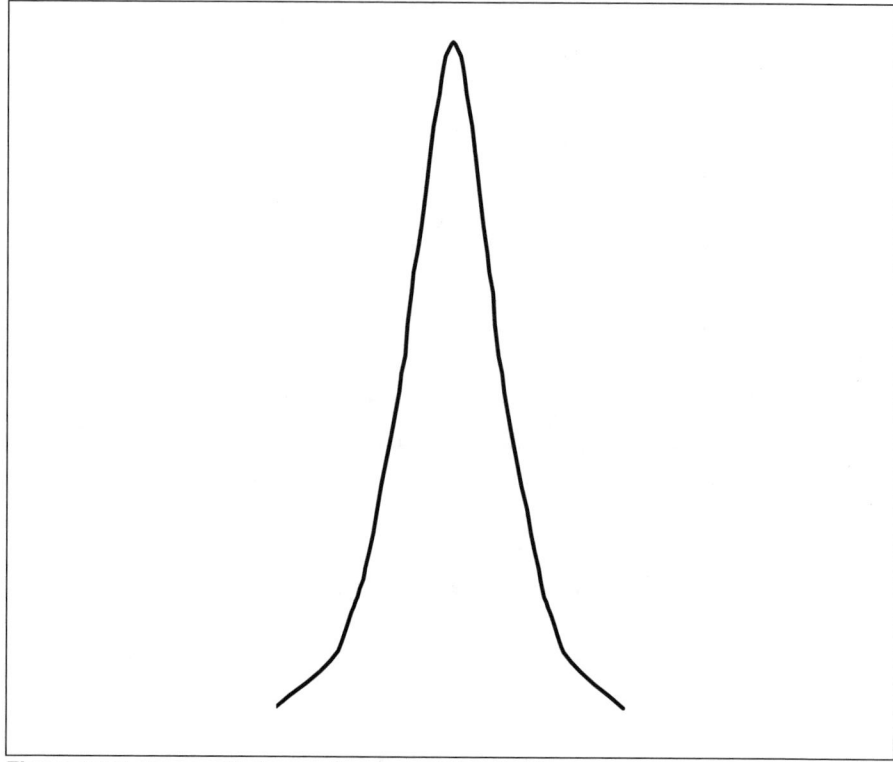

Figure 18-3. Narrow Variance Accuracy Distribution

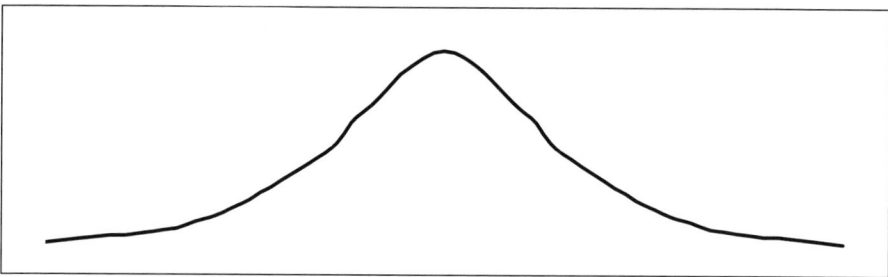

Figure 18-4. Wide Variance Accuracy Distribution

For the data shown in Table 18-1, the mean or average is zero since there are the same values above zero as there are below zero.

The variance is calculated by determining the mean, subtracting the mean from each value, squaring each result of the subtraction, and summing the squares. This sum is then divided by the number of samples minus 1. The result is the variance of the sample.

In the case of the data in Table 18-1 the variance is 0.0157.

If we take the square root of this number, we obtain another statistical function called the standard deviation. The value of this is 0.125 for the data in Table 18-1.

For our normal distribution this function, called the standard deviation, allows us to determine how many of the accuracy values will fall between certain limits. We can then say the same thing about any of the transducers manufactured since the sample is representative of the total population of transducers. This is true as long as the production methods, materials, and production machinery remain constant.

We can say that 68% of the accuracy values will fall between ±1 standard deviation or +0.125 and -0.125. We can also say that approximately 95% of the accuracy values will fall between ±2 standard deviations or +0.250 and -0.250. We can also say that approximately 99% of the accuracy values will fall between ± 3 standard deviations or +0.375 and -0.375.

By determining the variance and then the standard deviation, the manufacturer now has a means of finding out how many of the transducers will be likely to exceed a certain accuracy value. The manufacturer could publish an accuracy of ±0.25% of full scale and be certain that 95% of the transducers would meet this figure. A more conservative manufacturer would publish an accuracy of ±0.375% of full scale and be certain that 99% of the transducers would meet this figure.

Users should determine the standard deviation associated with an accuracy specification when selecting or comparing transducers. The normal practice is to quote an accuracy specification at a standard deviation of two (two standard deviations), but this is seldom stated in the specifications. The user may have to contact the manufacturer to verify the practice.

18.2 Root Sum Square Versus Sum of Errors

The question often arises of how to combine several error specifications or combine the accuracy specifications of several components in a measurement system to arrive at a single error or accuracy statement. The previous section showed that the accuracy specifications for a given component are based on a statistical analysis of a sample of several different components. This analysis also showed that the stated accuracy is simply a limit to the accuracy of a certain percentage of the components. There are many components with better accuracy and a few components with worse accuracy.

A measurement system could be comprised of a transducer, a signal conditioner, and a display. The transducer could have a stated accuracy of ±0.1% of full scale. The signal conditioner could have a stated accuracy of ±0.2% of full scale. The display could have a stated accuracy of ±0.25% of full scale.

The worst-case accuracy of the measurement system could be determined by simply adding these three accuracies to arrive at a system accuracy of ±0.55% of full scale. This is called the sum of the errors. This is a very conservative approach since the individual accuracies are unrelated and independent. It is also unlikely that the accuracies are all at their extreme values and in the same direction. If the accuracies were at their extreme values but in different directions, we could have any of the following combinations of accuracies:

Transducer	Conditioner	Display		System
+0.1%	+ 0.2%	+ 0.25%	=	+0.55%
+0.1%	+ 0.2%	- 0.25%	=	+0.05%
+0.1%	- 0.2%	- 0.25%	=	-0.35%
-0.1%	- 0.2%	+ 0.25%	=	-0.05%
-0.1%	+ 0.2%	+ 0.25%	=	+0.35%
-0.1%	- 0.2%	- 0.25%	=	-0.55%

The chance of having the accuracies occur in the directions that cause a low system accuracy of 0.05% is as unlikely as having them all be in one direction and cause a high system accuracy.

A more realistic approach is known as the root sum square (rss) determination. This is calculated by squaring each component's accuracy, summing the squares, and then taking the square root of the sum. The root sum square accuracy for the three component system is ± square root of $(0.1^2 + 0.2^2 + 0.25^2) = ±0.335\%$ of full scale. This value is a logical compromise between the low (±0.05%) and high (±0.55%) values obtained by summing the component errors as shown previously.

This root sum square calculation can also be used to determine the component's overall error from a number of individual component errors. Such individual errors could include hysteresis, linearity, and repeatability.

The units of the combined accuracies or errors must be the same before they can be used in a root sum square calculation. For example, the accuracies of the transducer, signal conditioner, and display must all be expressed as percentage of full scale or all as percentage of reading. The root sum square value is then expressed in the same units as the individual components.

The root sum square calculation using the basic accuracy values of the transducer, signal conditioner, and display will provide an initial estimate of the system's overall accuracy. If this calculated value does not meet or come close to the required measurement accuracy then it is pointless to continue a more in-depth analysis of the accuracy, as shown in section 19.1.

18.3 Standards Accuracy

The accuracy of the standards used to calibrate a transducer or measurement system must be included in the overall accuracy statement. This means that the accuracy of a calibrated transducer can never be better than the accuracy of the standards used in the calibration. It also means that the more accurate the standards are, the more accurate the calibrated transducer will be. The converse of this is that high-accuracy transducers will require very high-accuracy standards if the high level of measurement accuracy is to be maintained.

The accuracy of the standards can be included in the root sum square calculation to arrive at an overall system accuracy. Again, the accuracies must all be expressed in the same units before being used in the root sum square calculation.

The accuracy value provided by a transducer manufacturer may not include the accuracy of the standard. This is valid since the manufacturer is simply stating the extremes of accuracy for a certain percentage of the total number of transducers that are made. The value is not the accuracy of a specific transducer calibrated by a specific standard under specific conditions.

18.4 Test Uncertainty Ratio

The Test Uncertainty Ratio (TUR) or Test Accuracy Ratio (TAR) is a measure of how the accuracy of the standard compares to the accuracy of the transducer or system being calibrated. Many years ago, the accepted ratio was 10:1, or the accuracy of the standard was to be ten times better than the accuracy of the transducer or system. Nowadays, the accuracy of transducers has increased greatly while the accuracy of standards has not increased to the same degree. The accepted ratio is now 4:1, or the standard's accuracy is four times better than the transducer or system.

Consider the three cases when the ratios are 10:1, 4:1, and 1:1. The estimate of the effect of the standard's accuracy on the overall calibration accuracy can be calculated by using the root sum square approach.

Transducer accuracy: ± 1% of full scale
Standard accuracy: ± 0.1% of full scale
TAR   10:1

Overall accuracy: ± square root of $(1^2 + 0.1^2)$ = ± 1.005% of full scale

Transducer accuracy: ± 1% of full scale
Standard accuracy: ± 0.25% of full scale
TAR   4:1

Overall accuracy: ± square root of $(1^2 + 0.25^2)$ = ± 1.031% of full scale

Transducer accuracy: ± 1% of full scale
Standard accuracy: ± 1% of full scale
TAR   1:1

Overall accuracy: ± square root of $(1^2 + 1^2)$ = ± 1.414% of full scale

At a ratio of 10:1, the overall accuracy is only increased (made worse) by 0.5% compared to the accuracy of the transducer alone. At a ratio of 4:1, the overall accuracy is increased by 3.1%, while at a ratio of 1:1 the overall accuracy is increased by 41.4%.

18.5 Uncertainty Versus Accuracy

The accuracy of a transducer or measurement system is a specification of the instruments and does not tell us much about the accuracy of the measurement made with the instruments. The term *measurement uncertainty* is used to describe a calculated value that informs users of test results about how confident they can be that the measurement is within a stated amount above or below the true value. The uncertainty is necessary to help the user decide if the result is adequate for its intended purpose and to determine if it is consistent with other similar results.

The International Organization for Standardization's (ISO) *Guide to the Expression of Uncertainty in Measurement* defines uncertainty as "a parameter, associated with the results of a measurement that characterizes the dispersion of the values that could reasonably be attributed to the measurand." The measurand is the quantity being measured (pressure, temperature, flow, etc.). The ISO guide was produced to provide a standard method for expressing measurement uncertainty on a worldwide basis. National standards laboratories such as NIST in the United States and NRC in Canada have adopted this standard to describe the uncertainty of their own measurements. Government and commercial standards laboratories have also adopted the method, as have accreditation organizations that certify calibration laboratories. Commercial organizations and test laboratories are using the method on a more frequent basis when conducting measurements during their in-house activities.

The quality assurance program requirements for traceable calibrations also require that a statement of uncertainty accompany each calibration. This statement is used to support the unbroken chain of calibration back to the national or international standards. The uncertainty of the standard used in the calibration is included in the calculation of the uncertainty of the measurements made during the calibration. The uncertainty of the standard includes the uncertainty of the standards used in its calibration. The unbroken chain back to the national or international standard means that the uncertainty of each level of calibration includes the uncertainty of the level above it. This enables the user to define his or her measurement uncertainty in terms of the uncertainty of the national or international standard, which is, in effect, the true value.

The method of calculating the uncertainty of a measurement is not a simple procedure. The ISO guide is a 100-page document, while the NIST Technical Note 1297 on the subject is a 25-page document. There are standalone software programs and features in commercial calibration software that automate the procedure, but the user should be familiar with the details of the method before using the software.

The method for determining the uncertainty of a measurement starts with writing a measurement equation. This equation shows how the measurand (quantity being measured) Y is determined from N quantities $X_1, X_2, ..., X_N$ through a functional relation f rather than being measured directly. The equation then takes the form:

$$Y = f(X_1, X_2, ..., X_N)$$

The quantities Xi include correction factors and other sources of variability such as different instruments, samples, users, and times of the day. The function is a description of the measurement process and not just a physical law. The function should include all the quantities that could significantly affect the uncertainty of the measurement. Correction factors that are applied as part of the data reduction process are not included in this uncertainty analysis. Only the variable part of the correction factor would be included. For example, the specifications for the transmitter described in section 19.1 state that the effect of static pressure on the span is -1% ±0.25% of full scale per 1000 psig. The -1% effect is eliminated during calibration and thus is not part of the measurement equation. The ±0.25% effect is a variable that must be included in the equation.

The ISO guide gives the example of measuring the power dissipated in a resistor when a voltage is applied across the resistor. The resistor has a linear temperature coefficient of resistance that increases the resistance as the temperature rises. The resistor has a resistance of R_0 at temperature t_0, a temperature coefficient of resistance of b, and a voltage of V. The measurand, which is power (P) at the temperature t, depends on V, R_0, b, and t.

The physical law equation is:

$$P = V^2/R_0$$

The measurement equation is:

$$P = V^2/R_0 \, [1 + b \, (t - t_0)]$$

The measurement equation includes the correction factor caused by the temperature coefficient of resistance.

The estimate of the measurand is obtained by obtaining estimates of the N quantities and producing the following equation:

$$y = f(x_1, x_2, \ldots, x_N)$$

The uncertainty of the measurement depends on the uncertainties (u_i) of the estimates of the quantities (x_i). The different components of uncertainty are divided into two types based on the method that is used to evaluate the uncertainty.

Type A evaluations determine the uncertainty by statistical analysis of a series of observations. Type B evaluations determine the uncertainty by some means other than statistical analysis of a series of observations. Eventually, the two types of evaluations are combined into a single uncertainty.

Both types of uncertainties are represented by an estimated standard deviation that is called the standard uncertainty. This standard uncertainty (u_i) is equal to the positive square root of the estimated variance.

Type A uncertainties are determined by a statistically estimated standard deviation (s_i), which is equal to the positive square root of the estimated variance (s_i^2) along with the number of degrees of freedom.

For these uncertainties, $u_i = s_i$. The methods for calculating this variance and standard deviation are shown in section 18.1, earlier in this chapter.

Type A uncertainties could be the result of a series of measurements of the measurand taken under the same conditions. A type B uncertainty, represented by u_j, which is an approximation of the standard deviation, is equal to the positive square root of u_j^2, which is an approximation of the variance. This variance is obtained by assuming a probability distribution based on all of the available information. The method treats u_j and u_j^2 like statistically determined standard deviation and variance, respectively, and thus the standard uncertainty is u_j.

Type B uncertainties are based on information such as:

- Similar previous measurement data
- Experience and knowledge of how instruments, materials, and equipment operate in a similar environment
- Specifications provided by a manufacturer
- Calibration data and other test data for the instruments used
- Reference data uncertainties

These types of uncertainties are obtained from outside sources or from probability distributions that are assumed based on knowledge of the uncertainty.

For situations commonly found in practice, the estimates (x_i) are not correlated, and the measurement equations and the associated combined type A and type B standard uncertainties fall into one of the two following forms:

The measurement equation consists of a sum of quantities X_i multiplied by constants a_i.

$$Y = a_1 X_1 + a_2 X_2 + \ldots a_N X_N$$

The measurement result becomes:

$$y = a_1 x_1 + a_2 x_2 + \ldots a_N x_N$$

The combined standard uncertainty $[u_c(y)]$ becomes:

The square root of the estimated variance given by

$$u_c^2(y) = a_1^2 u^2(x_1) = a_2^2 u^2(x_2) + \ldots a_N^2 u^2(x_N)$$

The measurement equation consists of a product of quantities X_i raised to powers a, b, … p, all multiplied by a constant A:

$$Y = AX_1^a X_2^b \ldots X_N^p$$

The measurement result becomes:

$$y = Ax_1^a x_2^b \ldots x_N^p$$

The combined standard uncertainty $[u_{cr}(y)]$ becomes:

The square root of the estimated variance given by

$$u_{cr}^2(y) = a^2 u_r^2(x_1) + b^2 u_r^2(x_2) + \ldots p^2 u_r^2(x_N)$$

Where $u_{cr}(y)$ is the relative combined standard uncertainty of y.

When the probability distribution that is associated with a measurement y and its combined standard uncertainty is approximately normal and $u_c(y)$ is a good estimate of the standard deviation of y, then the interval from $y - u_c(y)$ to $y + u_c(y)$ will likely contain 68% of the values of the quantity Y of which y is an estimate. This also means that with a confidence level of 68%, Y is greater than or equal to $y - u_c(y)$ and less than or equal to $y + u_c(y)$. This is commonly written as $Y = y \pm u_c(y)$.

The measure of uncertainty that is commonly used in measurements is called the "expanded uncertainty U," which is obtained by multiplying $u_c(y)$ by a coverage factor k. This results in the equation $U = k\, u_c(y)$, where it is confidently determined that Y is greater than or equal to $y - U$ and is less than or equal to $y + U$. This is then written as $Y = y \pm U$.

The value used for the coverage factor k is based on the desired level of confidence that we want to associate with the interval defined by U = k u_c. The value of k is usually in the range of 2 to 3. For a normal distribution, k = 2 defines an interval that has a level of confidence of approximately 95%. A value of k = 3 defines an interval that has a confidence over 99%.

This description of the method for determining the measurement uncertainty should be considered as only a summary. The reader is referred to the ISO guide or the NIST Technical Note 1297 for a complete description.

19
Transducer Specifications

Selecting an appropriate transducer involves making a detailed analysis of the specifications provided by the manufacturer. This is done to ensure that the transducer meets the environmental test conditions, the measurement requirements, and the measurement error allowance. Environmental conditions can include temperature, pressure, humidity, vibration, radiation, and water resistance. Measurement requirements can include operating range, frequency response, space limitations, output signal compatibility, operating life, mounting requirements, and test fluid compatibility. The measurement error allowance provides input into the total measurement uncertainty analysis described in chapter 18, section 18.5, "Uncertainty Versus Accuracy." Before incorporating the selected transducer into a measurement system, a preliminary analysis of the potential error sources and their possible corrections should be performed prior to the uncertainty analysis to verify that the selected transducer alone has sufficient measurement accuracy. This analysis should be done before the transducer is purchased or before the final design of the measurement system.

Bench tests of the transducer's performance must be carried out for any conditions or combination of conditions not specified by the manufacturer. Similar tests are required if the transducer is modified or mounted in ways not specified by the manufacturer.

Reputable and successful transducer manufacturers will provide all the information available on their products and will usually be able to provide the test data to confirm their specifications. Suppliers that sell equipment made by another manufacturer under their own name may not be able to provide complete information. In this case, you may have to determine who the original manufacturer was and obtain the information from them.

19.1 Interpreting Manufacturer's Specification Sheets

The specification sheet provides detailed information about the environmental limits, environmental effects, operating limits, and installation effects. This data enables the user to determine if the equipment is suitable for the application and to determine the measurement accuracy under the installed environmental and mounting conditions.

The best way to illustrate how to interpret and use a specification sheet is to examine an actual sheet, calculate potential errors, and determine if the errors can be eliminated or reduced. The selected instrument is a differential pressure transmitter made by a well-known and reputable North American manufacturer. This is a process type instrument that is also widely used in test measurement activities. The product is an extremely popular instrument with an installed base of over four million units since its introduction in 1969. The specification sheet is very detailed and provides all the information required to calculate how the instrument will perform under operating conditions.

When introduced in 1969, this instrument was the first of its type to specify an accuracy of ±0.2% of calibrated span. The question is whether the user can calibrate this instrument in a calibration laboratory, install it, and expect to make measurements with this same accuracy. The process of using and interpreting the specification sheet will show that the user will have to perform a large number of steps to get anywhere near this accuracy figure. If the user fails to complete all the steps, he or she may not be aware that the measurement error is much greater than the advertised accuracy of ± 0.2%.

The differential pressure transmitter is designed to measure the difference in pressure between two locations in a test rig or process. It is capable of measuring very small pressure differences in the presence of very high static or background pressure. Typical uses include pressure drops across flow elements such as orifice plates or venturis and fluid level measurements in pressurized tanks. A typical model can be calibrated to measure an orifice plate differential pressure or tank level of 0 to 30 inches of water in a process that is operating at a static pressure of 1500 psig. The instrument is usually connected to the process equipment using stainless steel tubing. The tubing allows the instrument to be installed at a convenient location and remote from any high process fluid temperatures. A special valve manifold is used as part of the installation. This valve arrangement is used to eliminate some of the errors found during the analysis of the specifications. The operating principle of the instrument is based on the capacitive pressure transducer described in chapter 4, section 4.2.5. "Capacitive Pressure Transducers."

Figure 19-1 shows the transmitter measuring the differential pressure across an orifice plate caused by the flow in the pipe. The static pressure of the fluid in the pipe is 1500 psig. The differential pressure at the maximum flow is 30 inches of water. Valves A, B, and C are used to eliminate some of the errors caused by the operating environment and conditions. These valves are often incorporated into a single manifold block called an "equalizing manifold."

The differential pressure transmitter selected for this application has an upper range limit (URL) of 150 inches of water and a rangeability of 6:1. This means that the maximum possible calibrated range is from 0 to 150 inches of water, and the minimum possible calibrated range is from 0 to (150/6) = 25 inches of water. For this application, the transmitter is calibrated with a range of 0 to 30 inches of water. This range of 0 to 30 inches of water is equal to a calibrated span of (30 − 0) = 30 inches of water.

The instrument is calibrated in a calibration laboratory at a room temperature of 20°C. The calibration is carried out with the low side of the transmitter open to atmospheric pressure and the calibration source pressure connected to the high side. This results in a static pressure of 0 psig. The transmitter is calibrated while it is mounted vertically, and the calibration is repeated after a 12-month recalibration period.

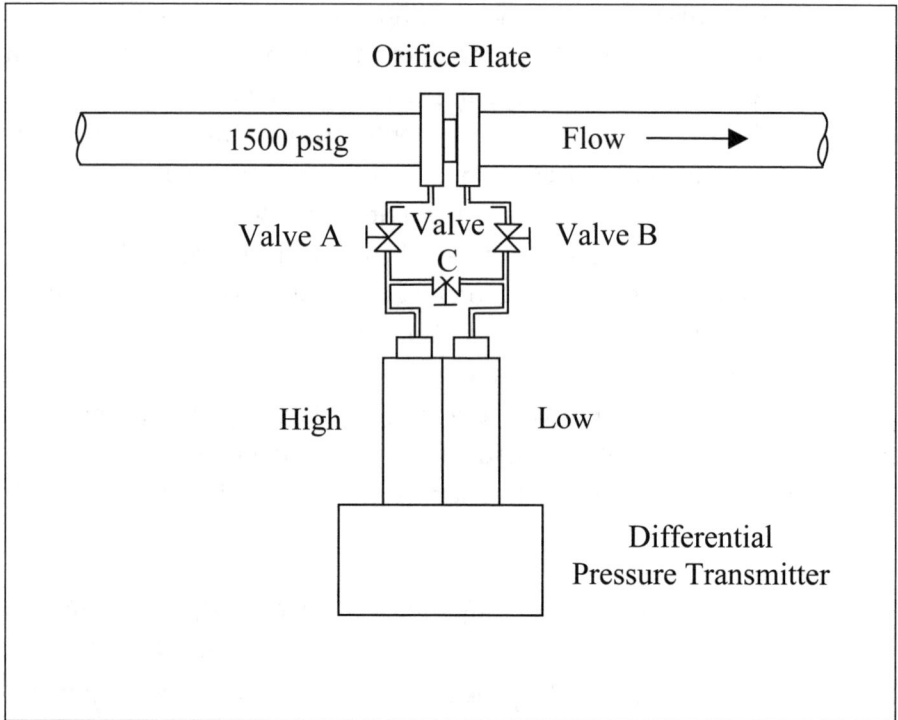

Figure 19-1. Differential Pressure Transmitter Installation

The transmitter is installed at a location where the ambient temperature is as high as 40°C on a process that has a static pressure of 1500 psig. There is the possibility that the transmitter will be mounted horizontally when installed.

The specification sheet should be examined for stated errors that will occur as a result of differences between the calibration conditions and the installed conditions.

The following error specifications are involved in this analysis:

- Accuracy Effect
- Temperature Effect
- Static Pressure Effect
- Stability Effect
- Mounting Effect

The following are the manufacturer's specifications for these effects:

    a. Accuracy (includes hysteresis and linearity)

        ±0.2% of calibrated span

    b. Temperature Effect

        Zero effect
        ±0.5% of URL per 56°C

        Span effect
        ±0.5% of calibrated span per 56°C

    c. Static Pressure Effect

        Zero effect
        ±0.25% of URL per 2000 psig

        Span effect
        -1% ±0.25% of reading per 1000 psig

    d. Stability effect

        ±0.25% of URL per 6 months

    e. Mounting effect

        Zero effect
        Up to ±1 inch of water

These specifications show that the effects are stated in a number of different ways. Some effects are stated as a percentage of the calibrated span, some are stated as a percentage of reading, and some are stated as a percentage of the upper range limit (URL). In some cases, the specifications show that effects can have one action on the zero value and another action on the span.

The static pressure effect on the span is stated as -1% ±0.25% of reading per 1000 psig. This means that there is a known effect of -1% of reading per 1000 psig and an unknown effect of up to ±0.25% of reading per 1000 psig. The known effect can be eliminated during calibration for a given static pressure. The unknown effect can only be determined by calibrating the instrument at the given static pressure.

In order to calculate an overall error, the effects must all be stated in consistent units. The simplest unit to use in this case is inches of water.

   a. Accuracy Effect (includes hysteresis and linearity)*

$\pm 0.2\%$ of calibrated span
$= \pm (0.2 / 100) \times 30 = \pm 0.06$ inches of water

* Note that the manufacturer has included hysteresis and linearity errors in the accuracy specification. Other manufacturers may not do this, so the user would have to add the errors as part of the analysis.

   b. Temperature Effect

Zero effect
$\pm 0.5\%$ of URL per 56°C

For $(40° - 20°) = 20°C$ change from calibrated temperature to in-use temperature:

Error is $\pm (20 / 56) \times (0.5 / 100) \times 150$ inches
$= \pm 0.27$ inches of water

Span effect
$\pm 0.5\%$ of calibrated span per 56°C

For $40° - 20° = 20°C$ change from calibrated temperature to in-use temperature:

Error is $\pm (20 / 56) \times (0.5 / 100) \times 30$ inches
$= \pm 0.045$ inches of water

   c. Static Pressure Effect

Zero effect
$\pm 0.25\%$ of URL per 2000 psig

At 1500 psig error is:

$\pm (1500/2000) \times (0.25/100) \times 150$ inches of water
$= \pm 0.28$ inches of water

Span effect
-1% ± 0.25% of reading per 1000 psig

At a reading of 30 inches of water and a static pressure of 1500 psig, error is:

$- (1/100) \times (1500/1000) \times 30 \pm (0.25/100) \times (1500/1000) \times 30$ inches of water

$= -0.3 \pm 0.11$ inches of water

At a reading of 5 inches of water and a static pressure of 1500 psig, error is:

$- (1/100) \times (1500/1000) \times 5 \pm (0.25/100) \times (1500/1000) \times 5$ inches of water

$= -0.05 \pm 0.02$ inches of water

d. Stability Effect

± 0.25% of URL per 6 months

Since unit is calibrated annually, the effect is doubled.

Error is $\pm 2 \times (0.25/100) \times 150$ inches of water
$= \pm 0.75$ inches of water

e. Mounting Effect

± 1 inch of water

This maximum effect would occur if the transmitter was calibrated in a vertical position and then installed in a horizontal position. The error is due to the difference in the head of water on both sides of the transmitter diaphragm, which is approximately one-inch thick. The error would be positive if the high side was on the bottom and negative if the low side was on the bottom.

Total error at a reading of 30 inches of water:

Accuracy:	± 0.06 inches of water
Temperature effect:	± (0.27 + 0.045) inches of water
	= ± 0.315 inches of water

Static pressure effect: $\pm\, 0.28 - 0.3 \pm 0.11$ inches of water
$= -0.3 \pm 0.39$ inches of water

Stability effect: $\pm\, 0.75$ inches of water

Mounting effect: $\pm\, 1$ inch of water

Root sum square of errors (except -0.3 inches known bias on the span due to static pressure):

$= \pm$ square root of $(0.06^2 + 0.315^2 + 0.39^2 + 0.75^2 + 1^2)$

$= \pm$ square root of $(0.0036 + 0.0992 + 0.1521 + 0.5625 + 1)$

$= \pm\, 1.35$

Add -0.3 inches known bias:

Total error at 30 inches of water is $+1.05, -1.65$ inches of water.

As a percentage of full scale (30 inches) this is an error of
$+ (1.05/30) \times 100 = +3.50\,\%$

$- (1.65/30) \times 100 = -5.50\,\%$

Total error at a reading of 5 inches of water:

Accuracy: $\pm\, 0.06$ inches of water

Temperature effect: $\pm\, (0.27 + 0.045)$ inches of water
$= \pm\, 0.315$ inches of water

Static pressure effect: $\pm\, 0.28 - 0.05 \pm 0.02$ inches of water
$= -0.05 \pm 0.30$ inches of water

Stability effect: $\pm\, 0.75$ inches of water

Mounting effect: $\pm\, 1$ inch of water

Root sum square of errors (except -0.05 inches known bias)

$= \pm$ square root of $(0.06^2 + 0.315^2 + 0.30^2 + 0.75^2 + 1^2)$

$= \pm$ square root of $(0.0036 + 0.0992 + 0.09 + 0.5625 + 1)$

$= \pm\, 1.32$

Add -0.05 inches known bias:

> Total error at 5 inches of water is + 1.27, - 1.37 inches of water.

As a percentage of reading (5 inches) this is an error of:

> + (1.27/5) × 100, - (1.37/5) × 100 = + 25.4, -27.4 % of reading

By adjusting the zero of the transmitter once it is installed, temperature stabilized, and pressurized to the static pressure we can eliminate any of the zero error effects. These errors include the zero effect due to static pressure, the zero effect due to temperature, and the mounting effect. The zero reading at static pressure is obtained by using the equalizing valve system (see Figure 19-1). The normal valve operation is to have valves A and B open and valve C closed. This allows the transmitter to measure the pressure drop across the orifice that occurs due to the flow. By closing valve B and opening valve C we can apply equal static pressure to both sides of the transmitter. Under these conditions, the transmitter output can be set to its zero value using the zero adjustment potentiometer. The -1% of reading per 1000 psig span effect error can also be eliminated by compensation during bench calibration.

This leaves us with the following errors:

   a. Accuracy Effect

   ± 0.2% of calibrated span

   = ± (0.2 / 100) × 30 = ± 0.06 inches of water

   b. Temperature

   Span effect
   ± 0.5% of calibrated span per 56°C

   For (40° − 20°) = 20°C change from calibrated temperature to in-use temperature:

   Error is ± (20 / 56) × (0.5 / 100) × 30 inches
   = ± 0.045 inches of water

c. Static Pressure

   Span effect
   ± 0.25% of reading per 1000 psig

   At a reading of 30 inches of water and a static pressure of 1500 psig, error is:

   ± (0.25/100) × (1500/1000) × 30 inches of water
   = ± 0.11 inches of water

   At a reading of 5 inches of water and a static pressure of 1500 psig, error is:

   ± (0.25/100) × (1500/1000) × 5 inches of water
   = ± 0.02 inches of water

d. Stability effect

   ± 0.25% of URL per 6 months

   Since unit is calibrated annually the effect is doubled.

   Error is ± 2 × (0.25/100) × 150 inches of water
   = ± 0.75 inches of water

Total error at a reading of 30 inches of water:

Accuracy:	± 0.06 inches of water
Temperature effect:	± 0.045 inches of water
Static pressure effect:	± 0.11 inches of water
Stability effect:	± 0.75 inches of water

Root sum square of errors

= ± square root of (0.0036 + 0.0020 + 0.0121 + 0.5625)
= ± 0.76 inches of water

Total error at 30 inches of water is ± 0.76 inches of water.

As a percentage of full scale (30 inches) this is an error of:

± (0.76/30) × 100 = ± 2.53 %.

Total error at a reading of 5 inches of water:

    Accuracy: ± 0.06 inches of water

    Temperature effect: ± 0.045 inches of water

    Static pressure effect: ± 0.02 inches of water

    Stability effect: ± 0.75 inches of water

Root sum square of errors

    = ± square root of (0.0036 + 0.0020 + 0.0040 + 0.5625)

    = ± 0.76 inches of water

Total error at 5 inches of water is ± 0.76 inches of water.

As a percentage of reading (5 inches) this is an error of:

    ± (0.76/5) × 100 = ± 15.2 %.

The major contributor to the remaining error is the stability effect. For this particular transmitter the error can only be reduced by calibrating at more frequent intervals. By calibrating every three months instead of once per year the stability error can be reduced to 25% of the original error.

Total error at a reading of 30 inches of water:

    Accuracy: ± 0.06 inches of water

    Temperature effect: ± 0.045 inches of water

    Static pressure effect: ± 0.11 inches of water

    Stability effect: ± 0.19 inches of water

Root sum square of errors

    = ± square root of (0.0036 + 0.0020 + 0.0121 + 0.0361)

    = ± 0.23 inches of water

Total error at 30 inches of water is ± 0.23 inches of water.

As a percentage of full scale (30 inches) this is an error of:

$$\pm (0.23/30) \times 100 = \pm 0.77\,\%.$$

Total error at a reading of 5 inches of water:

Accuracy:	± 0.06 inches of water
Temperature effect:	± 0.045 inches of water
Static pressure effect:	± 0.02 inches of water
Stability effect:	± 0.19 inches of water

Root sum square of errors

$$= \pm \text{ square root of } (0.0036 + 0.0020 + 0.0040 + 0.0361)$$

$$= \pm 0.23 \text{ inches of water}$$

Total error at 5 inches of water is ± 0.23 inches of water.

As a percentage reading (5 inches) this is an error of:

$$\pm (0.23/5) \times 100 = \pm 4.6\,\%.$$

The variable portion (±0.25% of reading per 1000 psig) of the static pressure effect on the span can only be eliminated if a calibration at the high static pressure can be performed. This is possible with special dead-weight tester systems, but the cost of the equipment is usually too high to justify it. The temperature effect on the span can be eliminated if the transmitter can be maintained in a field-mounted enclosure that keeps the transmitter at a controlled temperature. The calibration would then have to be carried out at this controlled temperature.

The stability effect can also be reduced for this manufacturer's transmitter by selecting a model that has an upper range limit (URL) that is closer to the calibrated range. This error is specified as a percentage of the URL rather than as a percentage of the calibrated span. By selecting a model that has a URL of 30 inches of water rather than 150 inches of water, the error is reduced by a factor of (150/30) = 6.

The zero effects that were eliminated by adjusting the zero of the transmitter once it was installed are only valid for the conditions that existed at the time of adjustment. If the static pressure or the ambient temperature changes then the zero will be affected. These changes can be monitored and adjustments made using the data acquisition software. Newer versions of this transmitter have built-in sensors and a microprocessor that enable users to make these adjustments in real time.

The error analysis shows that the user must be aware of the complete instrument specifications and must make the necessary adjustments to reduce the overall installed error. It also shows that the actual measurement error is often greater than the manufacturer's advertised accuracy error alone.

19.2 Transducer Type

The description of the transducer type is a fundamental part of the instrument specifications. The description expands on the basic transducer identification to ensure that the appropriate transducer is selected for the user's application.

An example of a basic transducer description is "pressure transducer." This can be expanded to describe it as a gauge pressure transducer, an absolute pressure transducer, or a vacuum pressure transducer. A differential pressure transducer can further be described as a wet-wet, wet-dry, or dry-dry transducer depending on whether each side of the diaphragm can be exposed to liquids or not. A dynamic pressure transducer is capable of measuring pressure that is changing quickly over time. The user must determine if the response time or the frequency response is adequate for the test conditions. The user must also know if the dynamic pressure transducer is capable of static or DC measurements or if it has some low-frequency response cut-off value.

Another example of a basic transducer description is "temperature transducer." Further description could identify it as a thermocouple or an RTD. A thermocouple can be further described as a type J or iron-constantan thermocouple and as having an exposed junction or a sealed junction. An RTD is further identified by its resistance at the ice point (0°C), by its coil material (platinum, copper, etc.), and by its alpha coefficient.

19.3 Operating Principle

The description of the transducer's operating principle also provides information that helps users select the appropriate instrument. For example, a bonded strain gauge pressure transducer will not have as high a frequency response as a piezoelectric pressure transducer. It will also not likely have as high an ambient temperature limit as a the piezoelectric transducer. A laser-based displacement transducer will likely be more accurate, have better resolution, and have a higher frequency response compared to a potentiometer-based displacement transducer.

19.4 Accuracy (% of full scale vs. % of reading)

The accuracy specification of a transducer is probably the most used and most abused instrument specification. The user needs to know whether the selected transducer has sufficient accuracy over the range and operating conditions encountered during the test. To completely answer this question, the user must perform an analysis of the complete specifications as shown in section 19.1 of this chapter and an uncertainty analysis as described in chapter 18, section 18.5, "Uncertainty Versus Accuracy." The analysis will also show whether the manufacturer has quoted an impressive accuracy specification in the advertising but hidden less desirable effects deep in the specification sheet.

The basic accuracy specification gives the user some idea as to whether the transducer will be accurate enough for a given test. If the test requirements are to determine a parameter within ±0.1% then a transducer with an accuracy specification of ±1% will not suffice. It is likely that a transducer with an accuracy specification of ±0.1% will also not suffice because of the other specifications and installed conditions that must be accounted for. The user should probably select a transducer that has an accuracy specification that is 4 to 5 times better than the required measurement accuracy of the parameter.

An important distinction in accuracy specifications is knowing whether the accuracy is specified as a percentage of full scale or as a percentage of reading. Most manufacturers will specify accuracy as a percentage of full scale since it results in a lower absolute number. The user must determine whether this accuracy specification meets his or her test requirements. The differential pressure transmitter described in section 19.1 of this chapter has an accuracy specification of ±0.2% of calibrated span, which for a 0 to 30 inches of water range is the same as ±0.2% of full scale. This accuracy can also be expressed in terms of the measured parameter as ±0.06 inches of water. At a reading of 30 inches of water this represents a measurement error of 0.2% of reading. At a reading of 5 inches of water this represents a measurement error of $(0.06/5) \times 100 = 1.2\%$ of reading. The typical test requirement is to measure the parameter over a wide range with the same measurement accuracy. This means that the transducer must have its accuracy specified as a percentage of reading or have a much higher percentage of full scale accuracy specification.

The accuracy specification given in the specification sheet is based on the manufacturer's experience in calibrating a large number of transducers. Assuming a normal distribution in these results, some transducers will not meet the accuracy specification. A large number of transducers will also have a better accuracy than the specification. This can only be determined by calibrating the individual transducer prior to using it in a test.

19.5 Linearity

Linearity is an important specification for test measurement transducers since the instrument is typically used to make measurements over its entire range. This contrasts with process instruments, which are typically used to measure values over a small part of their range. Good linearity produces the same level of measurement accuracy regardless of the actual reading value. The output of a highly linear transducer can also be amplified to use a smaller part of its range without greatly increasing the measurement error.

19.6 Hysteresis

The hysteresis specification and the linearity specification may be included in the accuracy specification, or they may be stated separately. Hysteresis is particularly important in vibration transducers where the measurement direction is constantly oscillating about some mean value. The output signal will then differ depending on whether the input is increasing or decreasing.

19.7 Environmental Operating Limits

The transducer specifications state the limits of the environmental conditions under which the transducer can operate and remain undamaged. The manufacturer will not guarantee the operation or survival of the transducer if these limits are exceeded. The manufacturer should state if any of the limits are mutually inclusive. This means that a transducer may have a temperature upper-limit specification of 200°F and a case pressure upper-limit specification of 1000 psig. We assume that the transducer can operate in an environment of 200° and 1000 psig. If the transducer pressure limit is only 500 psig when the temperature is 200° then the manufacturer should state this fact.

The manufacturer may qualify these limits in a number of different ways. There may be different limits for operating and storage environments. There may be limits for which the calibration is not affected or is only affected as long as the limit exists. Other limits will guarantee the survival of the transducer but not guarantee the calibration.

19.7.1 Temperature
Temperature limit specifications include maximum and minimum values for both storage and operation. They may also include the maximum rate of change of temperature. The limit for the ambient temperature may be different from the limit on fluid temperature, although in most cases these will tend to equalize.

Temperature can affect built-in electronic components, material strength, insulation, bonding materials, and seal integrity. It can also affect metal spring rates and cause distortion due to different coefficients of expansion.

19.7.2 Pressure

Pressure specifications can include maximum and minimum input pressure ranges, maximum and minimum static or case pressures, maximum burst pressure, and maximum submersion depth for watertight transducers.

Pressure transducers that are designed for positive pressure measurement only may be damaged if a vacuum is applied to the input. Similarly, transducers that are designed to measure vacuum may be damaged if positive pressure is applied to the input.

The submersion depth specification is really a pressure limit since it defines the point at which the external fluid pressure will cause leakage into the body of the transducer. Transducers that are designed to be immersed in high-pressure fluids have welded or hermetically sealed joints and integral mineral-insulated, stainless steel sheathed cable.

19.7.3 Humidity

The humidity specification defines the ambient humidity range over which the transducer can operate or be stored. The storage range may often be wider than the operating range.

A typical humidity specification would state that the transducer could be operated in an ambient relative humidity from 0% to 95%, noncondensing. The qualifying statement "noncondensing" means that the transducer can tolerate this humidity range as long as the transducer body's temperature is not cold enough to cause internal or external condensation of the water vapor into a liquid. Another way of stating this is that the humidity range is from 0% to 95% as long as the transducer body's temperature is above the dew point.

19.7.4 Vibration

Excessive vibration of a transducer can cause errors in the output signal and can damage the transducer or its cable connection. For a transducer that is not designed to measure vibration, the manufacturer may specify a maximum vibration limit that will prevent errors and another limit that will prevent damage. The specification will include the direction of vibration (which transducer axis), the level of vibration (displacement, velocity, or acceleration), and the frequency range or limit.

For transducers that are designed to measure vibration, the manufacturer may also include the effect on the output signal for vibration in axes other than the measurement axis. The manufacturer may also describe any resonant frequencies and their peak values.

19.7.5 Side Loading

Side loading is a specification that manufacturers often provide for load cells. It defines the limit to any load applied to the transducer axis at right angles to the measurement axis. The specification may also define the effect on the output signal per unit of side load.

Excessive side loading can damage the transducer's mounting threads, damage internal components, or even distort the transducer body.

20
Test Procedure Design

Designing a comprehensive and successful test procedure requires a great deal of input from the test measurement professional. Laboratory tests are expensive to design, assemble, instrument, and carry out. Often there is only one chance to obtain the required measurements before part of the test equipment is altered or destroyed by the test.

It is almost always less expensive to spend time and money on a thorough preparation than to repeat a test because one or more items were not considered when the test was planned. Such preparation should entail gaining input from all sections of the group involved in the project. This can include design engineering, management, quality assurance, test rig designers, test measurement professionals, and test operators. All these sections should meet and come to a complete understanding on the test requirements, test procedures, and test design. Additional meetings between participating sections should be held to complete the details of the test design. Nothing is more counterproductive than to have a test rig designer finish constructing the rig without first consulting the test measurement staff. In this case, the test instrumentation would have to be jury-rigged onto the test equipment in an inefficient and suboptimum manner. There is often no space allocated to install instrumentation or to route wires and cables.

The following sections discuss information that the test measurement professional is responsible for understanding and inputting into the test design process.

20.1 Test Objective

The test objective is a written statement that completely describes the purpose of the test. It should include background information about why the test is being carried out, the range of test conditions that will be covered, and the format of the expected test results. It should also define any national or international standards that have to be met as well as any quality assurance requirements. Any relevant regulatory requirements should also be defined in the test objective.

The test objective is typically prepared by the design engineering staff with input from management and quality assurance staff. All the sections that will be involved in the test should review the test objective as early as possible.

20.2 Required Measurements

A description of the required measurements should be a natural byproduct of the preparation of the test objective. This description should include the measurement range for each test parameter. An initial description is typically prepared by the design engineering staff. It should then be reviewed by the test measurement professionals and the test rig operators. These people have the knowledge to determine if such measurements are possible and an idea of their approximate cost. They should also be able to describe any additional measurements that may need to be taken to support the basic required measurements. The test schedule requirements can have an impact on which measurements should be made. The delivery time for some test instrumentation can be several weeks or months, especially if some customization is required.

Before a final decision is made on the required measurements, testing professionals may need to consider the information described in the following ten sections.

20.3 Accuracy Requirements

Defining the accuracy requirements for each measurement is one of the most difficult tasks to resolve in the development of the test procedure. It is also one of the most important tasks because it will determine the test's success and have a large impact on its cost. If you underestimate the accuracy requirements then the test results may not support your conclusions about the test's objective. If you overestimate the accuracy requirements then the cost of instrumentation will be excessive.

The accuracy requirement should be provided by the design engineering staff since they should know what accuracy is required to satisfy the test objective. A response such as "as accurate as possible" or "the same as last time" means that the staff has not done their homework.

The test measurement professionals should be able to say whether the expected measurement accuracies can be met and at what cost.

20.4 Frequency Response Requirements

The frequency response requirements are essential for any dynamic measurements. They can include upper and lower frequency limits or time constants and response times. The selected instruments must be able to provide the dynamic measurements to the required accuracy over the range of test frequencies.

The phase shifts of transducer systems should also be considered if any frequency response type testing is being performed. The frequencies encountered during the test can also have an impact on the method used to mount the transducer. This can include accelerometer mounting stiffness and the length and diameter of pressure transducer's tubing. In some situations, the diaphragm of the pressure transducer has to be mounted flush with the inside wall of a cavity to prevent pressure oscillations from distorting the measurement signal.

The frequency response limits will also affect the sampling rate requirements of any data acquisition systems that are used to acquire test data. Remember the Nyquist theorem referred to in section 7.5, "Filters," which stated that the sampling frequency had to be at least two times greater than the highest frequency that occurred in the test data.

20.5 Transducer Test Environment Limits

Testing professionals must know the extremes in the transducer environment to ensure that the transducer is not damaged and to determine the effects of these extremes on measurement accuracy. All of the conditions described in section 19.7, "Environmental Operating Limits," should be evaluated before the instrumentation is selected. The normal test conditions as well as any potentially damaging conditions caused by test malfunctions should be included in this assessment.

It is possible that special installation designs may be needed to protect the transducer while it is installed on the test rig. These can include insulation, cooling, vibration isolation, and waterproofing. In some cases, a special enclosure may have to be constructed to protect the transducer from temperature, pressure, or radiation.

20.6 Transducer Connections and Space Limitations

In most test situations the process of installing the transducer and its cabling must affect neither the operation of the equipment being tested nor the measurement signal. The transducer's physical size, shape, and weight can often affect the response of the object on which it is mounted. The size, weight, and stiffness of the cabling can also cause problems.

The weight of the transducer can alter the vibration and frequency response of objects. The transducer's size and shape can also alter the flow distribution characteristics of fluids passing over the transducer and its cabling. The transducer can also influence the temperature of equipment through conduction or radiation.

The design of a pressure tap or the length and diameter of the tubing used to connect a pressure transducer to the test equipment can produce pressure oscillations in the volume next to the pressure transducer diaphragm. Each specific volume will resonate at its natural frequency, which is determined by the length, diameter, and speed of sound in the fluid. The frequency is similar to the frequency of an organ pipe.

Burrs on the inside of a pipe wall, which are created by drilling a pressure tap, can also create pressure oscillations as vortices are created by the fluid flowing over the burr. These burrs should be removed before installing the transducers. Situations such as this underscore how important it is to keep the instrumentation in mind when designing and constructing a test rig. It will often be difficult or impossible to remove such things as burrs after the test rig components are assembled.

Sometimes the transducer will have to be built into the piece of equipment that is being tested to maintain the equipment's shape or surface profile. This is often done with the pressure transducers and accelerometers that are attached to aircraft airfoils being tested in a wind tunnel.

20.7 Test Data Format

The form in which the test measurement data is to be acquired, displayed, and stored in must be decided upon and documented when the test procedures are developed. A simple test could be conducted in which a few transducers are displayed on digital indicators. When the test operator simply notes readings at appropriate times the form of data acquisition and storage is manual.

A slightly more sophisticated system would have indicators that stored the digital data at preset times. The data would be downloaded after the test was completed. In more sophisticated tests, a large number of transducers would be connected to a data acquisition system that would be programmed to acquire the data at predetermined times, scale the voltages into engineering units, and store the data. At the same time, a monitor would display the current values of all or selected transducers. The test operator would use this information to set and maintain test conditions as well as gather test results. Even more sophisticated systems would have a data acquisition system and a real-time data analysis program that processed the test data and performed some form of real-time analysis. Vibration analysis systems and strain gauge analysis systems are typical examples of this type of arrangement.

20.8 Test Data Analysis

Test data analysis can be carried out in real time as the test is occurring or after the test sequence is completed (post-test). It is also possible to combine the real-time and post-test analysis methods so that some analysis is done while the test is occurring and other analysis is carried out on the stored data after the test.

The factors that can influence the analysis method selected include:

- Duration of the test
- Availability of online analysis equipment
- Need to vary test parameters as results become available
- Need to carry out multiple types of analysis on the same data
- Requirement that analysis be done by people other than the test operators

The ideal situation is to store all raw test data that is acquired during the test. This allows unlimited post-test analysis and makes it possible to correct any errors that are discovered as part of this analysis.

The amount of data analysis that is to be performed should be carefully considered, particularly if graphs or plots are required. Even if the analysis software automates the graphing process it can take an extremely long time to produce the plots.

Consider a test in which six accelerometers are installed and the test involves measuring the vibration at five different flow rates and three different fluid temperatures. The design engineer requests a time domain plot for each condition and frequency spectra plots over two different frequency ranges for each condition. While this doesn't appear to be a large amount of data reduction, consider that we have 6 sensors × 5 flows × 3 temperatures × 3 plots for each. The total number of plots required is 270. In a case like this, it may be more efficient to first plot three flow and two temperature conditions and look at the data to see if there are trends or great differences between the results for these conditions. Based on this initial analysis, a decision can be made whether to continue plotting out all the data. This also shows the importance of real-time analysis that can identify trends or anomalies in the test data as the test is occurring. It also shows that having design personnel present during the test can potentially reduce the amount of data reduction needed since they could also see results in real time.

20.9 Transducer Selection

The selection of the appropriate transducers for a test is based on a review of all the factors considered in sections 20.1 to 20.6 as well as an analysis of the transducer specifications outlined in chapter 19.

In some cases, the transducer selected will be a compromise reflecting budget restraints, long deliveries, or extreme test conditions. Such compromises must be carefully evaluated to verify that they do not affect the objective of the test.

One requirement may be that a particular transducer's performance must be verified before it is used in a test. This can occur if the manufacturer's specifications do not state the effect of a certain test condition, if the transducer installation is somewhat abnormal, if the transducer has been modified, or if the transducer is being used beyond its published limits. This verification takes the form of a bench test that determines the transducer's response to the expected test conditions.

20.10 Calibration Requirements

The calibration requirements for all the instrumentation to be used in a test must be evaluated as part of the test procedure design. The calibration requirements can affect the transducer selection, the test budget, the test schedule, and the data analysis.

A decision has to be made as to when and how often the instrumentation must be calibrated. This determination may be dictated by regulatory requirements, national or international standards, quality assurance procedures, length of the test, severity of the test conditions, manufacturer's recommendations, and engineers' experience with the particular instrumentation. When components are pressure tested the pressure measurement instrumentation must usually be calibrated within a few days of the start of the test or series of tests. Some equipment qualification tests require that the instrumentation be calibrated before and immediately after the test to verify that the calibration has not changed. Quality assurance procedures may state that the instrumentation has to be calibrated at least every six months.

The plant professionals must know which instrumentation is to be calibrated in house and which must be sent to a calibration facility. The turnaround time for outside calibrations can seriously affect the test's schedule. When purchasing new instrumentation a decision has to be made whether to have the vendor perform an initial calibration as part of the purchase price. The alternative is to calibrate the instrument in house before the test.

The test procedure should also state whether the calibrations are to be component calibrations in which each individual piece of instrumentation is calibrated separately or system calibrations in which the transducer, signal conditioning, and display components are calibrated together as a completely wired system.

A measurement uncertainty analysis should be performed on each measurement system used in the test. This will ensure that the systems meet the test objective's accuracy requirements. The analysis will also demonstrate that the calibration equipment and procedure are adequate for the specified measurement uncertainty.

Software that is developed as part of the data acquisition or data analysis functions should be validated before it is used in the test.

Appendix 1
Internet Links to Measurement Instrumentation Information

Organizations

A2LA (The American Association for Laboratory Accreditation)
http://www.a2la.org/

ASTM (American Society for Testing and Materials; now ASTM International)
http://www.astm.org

IEC (International Engineering Consortium)
http://www.iec.org/

IEEE (The Institute of Electrical and Electronic Engineers)
http://www.ieee.org/portal/site

ISA (The Instrumentation, Systems, and Automation Society)
http://www.isa.org
ISA Standards
http://www.isa.org/standards

NIST (National Institute of Standards and Technology)
http://www.nist.gov/

NRC (National Research Council)
http://www.nrc-cnrc.gc.ca/

Magazines

Cal Lab
http://www.callabmag.com/

Intech
http://www.isa.org/InTech

Sensors
http://www.sensorsmag.com/

Test and Measurement World
http://www.reed-electronics.com/tmworld/

Calculators

Air Density Calculator
http://wahiduddin.net/calc/density_altitude.htm

Boiling Point of Water Calculator
http://www.biggreenegg.com/boilingPoint.htm#elevation

Measurement Unit Converter
http://www.davidson.com.au/tools/convert/

Tolerance Calculator
http://www.callabmag.com/freetolerance.pdf

Uncertainty Calculator
http://www.callabmag.com/freeuncertainty.pdf

Unit of Measure Converter (converts units of pressure, temperature, length, mass & flow + gas properties)
http://www.dhinstruments.com/suppl/software/pcunitconv.htm

Vibration Calculator
http://www.wilcoxon.com/knowdesk_calculator.cfm

Calibration

Calibration Application Notes (registration required)
http://us.fluke.com/usen/support/appnotes/
default.htm?category=ap_ftp(flukeproducts)

Calibration Information On-Line
http://www.ecalibration.com/ecalibration/index.htm

Calibration Instrumentation Directory
http://www.ecalibration.com/ecalibration/
InstrumentationDirectory.asp

Calibration Laboratory Accreditation
http://www.transcat.com/CalServices/PDFs/17025WhitePaper.pdf

Calibration Management Software
http://us.fluke.com/usen/products/
METCALPlus.htm?catalog_name=FlukeUnitedStates
http://www.coolblue.com/coolblue/
http://www.dds-inc.com/
http://www.primetechpa.com/
http://www.calibration-software.com/

Calibration Organizations Directory
http://www.ecalibration.com/ecalibration/
OrganizationsDirectory.asp

Calibration Services Directory
http://www.ecalibration.com/ecalibration/ServicesDirectory.asp

Calibration Supplies Directory
http://www.ecalibration.com/ecalibration/SuppliesDirectory.asp

Dead-Weight Tester Information
http://www.mecheng.tcd.ie/~clyons/PresMes.pdf

Liquid Flow Calibration Facility
http://homepages.compuserve.de/DrREngel/div_eng.pdf

NIST Calibration Services
http://ts.nist.gov/ts/htdocs/230/233/calibrations/index.htm

Shunt Calibration of Strain Gauge Instrumentation
http://www.vishay.com/company/brands/measurements-group/guide/tn/tn514/514index.htm

Some Notes on Device Calibration
http://www.mme.tcd.ie/~clyons/calibrations.pdf

Stopwatch and Timer Calibrations
http://mpti.org/PDFs/960-12.pdf

Temperature Calibration
http://www.branom.com/literature/calibration.html

The Proving Ring
http://www.mel.nist.gov/div822/proving_ring.htm

TN-9 Technical Note Addressing Shunt Calibration Technique for Strain Gauge Load Cells
http://pcb.com/techsupport/docs/ftq/TN-9-0202_Shunt_Calibration.pdf

TN-15 Technical Note Addressing Dynamic Pressure Calibration
http://pcb.com/techsupport/docs/prs/TN-15-0205_Dynamic_Pressure_Calibration.pdf

Water Triple Point Cells
http://www.isotech.co.uk/water.html

Measurement Uncertainty

A2LA Guide for Estimating Measurement Uncertainty in Testing
http://www.a2la.org/guidance/est_mu_testing.pdf

A2LA Policy on Estimating Measurement Uncertainty for Testing Laboratories
http://www.a2la.org/policies/Testing_MU_policy.pdf

A2LA Policy on Measurement Traceability
http://www.a2la.org/policies/tracepol.pdf

An Application of the *Guide to Measurement Uncertainty*
http://www.fluke.ie/comx/applications/deavermsc00.pdf

Engineering Statistics Handbook
http://www.itl.nist.gov/div898/handbook/index.htm

Guide to Measurement Uncertainty Software Download
http://www.agilent.com/metrology/download5.shtml

Handbook of Statistical Practice
http://www.tufts.edu/%7Egdallal/LHSP.HTM

Measurement Uncertainty Book
Measurement Uncertainty, 3rd ed., by Ron Dieck
http://www.isa.org/measuncertainty

Online Statistics Book
http://davidmlane.com/hyperstat/index.html

Online Statistics Course
http://psych.rice.edu/online_stat/index.html

Uncertainty in Gas Turbine Measurements
http://www.barringer1.com/drbob-bio_files/Nov73.pdf

Vibration

Measuring Vibration
http://www.bksv.com/pdf/Measuring_Vibration.pdf

Vibration Measurement Links
http://www.wilcoxon.com/knowdesk_items.cfm/KDtype=TN
http://www.vibrationworld.com/

Vibration Nomograph
http://www.derritron.com/downloads/pdfs/
DerritronVibnomo.PDF

Vibration and Pressure Transducer General Literature (registration required)
http://www.endevco.com/resources/genliterature.php

Vibration and Pressure Transducer Technical Data (registration required)
http://www.endevco.com/resources/techdata.php

Vibration and Pressure Transducer Technical Papers (registration required)
http://www.endevco.com/resources/techpapers.php

Vibration and Shock Glossary
http://www.vibrationandshock.com/glossary.htm

Temperature
High Temperature Instrumentation
http://extenv.jpl.nasa.gov/presentations/High-Temperature_Components_&_Circuits.pdf

Table of Emissivity
http://www.monarchserver.com/TableofEmissivity.pdf
http://www.infrared-thermography.com/material.htm
http://ib.cnea.gov.ar/~experim2/Cosas/omega/emisivity.htm

Temperature Measurement – Thermocouple and RTD Tables
http://www.thermometricscorp.com/thermometricscorp/technical.html

Strain Gauges

Dynamic Strain Measurement
http://sem.hnrnet.com/PUBS-ArtDownLoad-DSTOC.htm

Strain Gauges--Back to Basics
http://sem.hnrnet.com/PUBS-ArtDownLoad-SGTOC.htm

Strain Gauge Technology
http://www.vishay.com/company/brands/measurements-group/guide/index.htm

Pressure

Pressure Fundamentals and Transmitter Selection
http://www.scanivalve.com/pdf/info_presstemphdbk.pdf

Pressure and Temperature Handbook
http://www.scanivalve.com/pdf/info_presstemphdbk.pdf

Time Lag of Pressure Systems
http://www.flowkinetics.com/

Flow

Fundamentals of Flow Metering
http://www.rosemount.com/document/pds/flowfund.pdf

Orifice Meter Accuracy
http://www.flowcontrolnetwork.com/PastIssues/Jan2003/1.asp

Force

Proving Rings
http://www.mel.nist.gov/div822/proving_ring.htm

Displacement

The Gauge Block Handbook
http://emtoolbox.nist.gov/Publications/NISTMonograph180.pdf

A Primer on Displacement Measuring Interferometers
http://zygo.com/products/zmi/dmiprim.pdf

Selecting Position Transducers
http://www.spaceagecontrol.com/selpt.htm#close

Data Acquisition

Choosing Data Acquisition Boards and Software
http://www.datx.com/images/pdfs/cdab.pdf

Computer Technology for Data Acquisition, Data Analysis, and Control
http://www.physics.mcmaster.ca/phy4d6/Lab/home.htm

Data Acquisition Fundamentals
http://digital.ni.com/devzone/conceptd.nsf/webmain/139DFA3645B29BE586256865004E742A/$File/AN007.pdf

Sensors, Signal Conditioning, and Data Acquisition
http://ccrma.stanford.edu/CCRMA/Courses/252/sensors/sensors.html

Transducers, Sensors, and Signal Conditioning
http://www.onesmartclick.com/engineering/transducers-signal-conditioning.html

Measurement

Instrumentation Application Notes
http://www.branom.com/literature.html

Measurements Course
http://gozips.uakron.edu/%7Edorfi/

Measurement Tech Tips
http://www.davidson.com.au/tools/techtips/

Power of Measurement Workshop
http://www.apcs.net.au/kn/kn20001/kn20001.html

Piezoelectric Transducers

Driving Long Cables
http://pcb.com/techsupport/tech_cables.php

General Piezoelectric Theory
http://pcb.com/techsupport/tech_gen.php

Introduction to Piezoelectric Accelerometers
http://pcb.com/techsupport/tech_accel.php

Chapter Appendix 1 – Internet Links to Measurement Instrumentation Information

Introduction to Piezoelectric Force Sensors
http://pcb.com/techsupport/tech_force.php

Introduction to Piezoelectric Industrial Accelerometers
http://pcb.com/techsupport/tech_indaccel.php

Introduction to Piezoelectric Pressure Sensors
http://pcb.com/techsupport/tech_pres.php

Introduction to Signal Conditioning for ICP® & Charge Piezoelectric Sensors
http://pcb.com/techsupport/tech_signal.php

Transducers

Miniature Sensors: When, Where, and Why We Should Use Them
http://www.entran.com/TechTipPart1.htm

Strain Gauge–Based Transducers: Their Design and Construction
http://www.davidson.com.au/products/strain/mg/technology/technotes/sgbt.pdf

TN-11 Calculating Rise Time Capabilities for Shock and Blast Measurements
http://pcb.com/techsupport/docs/pcb/TN-11-0904_Calculating_Rise_Time.pdf

TN-16 Technical Note Addressing the Use of Placebo Transducers
http://pcb.com/techsupport/docs/vib/TN-16-0305_Placebo_Transducers.pdf

The Transducer Directory
http://www.transducerdirectory.info/

Data Analysis

FFT Relationship Between Sampling Rate, Bandwidth, Resolution, etc.:
http://www.vibrationworld.com/AppNotes/Sampling.htm

The Fundamentals of Signal Analysis - Application Note 243
http://cp.literature.agilent.com/litweb/pdf/5952-8898E.pdf

A Refresher Course on Windowing and Measurements
http://www.home.agilent.com/upload/cmc_upload/All/
6C06DATAACQ_WINDOWS.pdf

Rotor Dynamics Measurement Techniques – Product Note Agilent 35670A-1
http://www.home.agilent.com/upload/cmc_upload/All/
6C06DATAACQ_WINDOWS.pdf

Signal Conditioning

Industrial Signal Conditioning, A Tutorial
http://www.dataforth.com/catalog/pdf/DTF-Tutorial.pdf

On-Line Courses

Dynamic Force, Pressure, and Acceleration Course http://tce.ing.uniroma1.it/personale/blasi/misurels/materiale/mat7/mat7a/dynamic-1.pdf
http://tce.ing.uniroma1.it/personale/blasi/misurels/materiale/mat7/mat7a/dynamic-2.pdf
http://tce.ing.uniroma1.it/personale/blasi/misurels/materiale/mat7/mat7a/dynamic-3.pdf
http://tce.ing.uniroma1.it/personale/blasi/misurels/materiale/mat7/mat7b/dynamic-4.pdf
http://tce.ing.uniroma1.it/personale/blasi/misurels/materiale/mat7/mat7b/dynamic-5.pdf
http://tce.ing.uniroma1.it/personale/blasi/misurels/materiale/mat7/mat7b/dynamic-6.pdf
http://tce.ing.uniroma1.it/personale/blasi/misurels/materiale/mat7/mat7b/dynamic-7.pdf
http://tce.ing.uniroma1.it/personale/blasi/misurels/materiale/mat8/dynamic-8.pdf
http://tce.ing.uniroma1.it/personale/blasi/misurels/materiale/mat8/dynamic-9.pdf
http://tce.ing.uniroma1.it/personale/blasi/misurels/materiale/mat8/dynamic-10.pdf

Experimental Methods – Course Manual
http://www.aoe.vt.edu/~devenpor/aoe3054/manual/index.html

Miscellaneous

International Vocabulary of Basic and General Terms in Metrology (VIM),
3rd ed. (draft) April 2004
http://www.cmi.cz/rod/222-IS-A.pdf

Programmable Logic Controllers (PLCs)
http://www.canadu.com/hjhtml/

Appendix 2
Internet Links to General Engineering Information

Conversion Factors for Energy Equivalents
http://physics.nist.gov/cuu/Constants/energy.html

Engineering Conversion Factors
http://www.processassociates.com/process/convert/cf_all.htm

Engineering Unit Converters/Calculators
http://www.edasolutions.com/Groups/Tech/EngineeringUC.htm

Fundamental Physical Constants
http://physics.nist.gov/cuu/Constants/index.html

List of Acronyms
http://www.edasolutions.com/acronyms/a_c.htm

Mass Flowrate Conversion Calculator
http://www.processassociates.com/process/convert/cf_flm.htm

On-Line Physics Textbook – Vibrations and Waves, Electricity and Magnetism, Optics and More
http://www.lightandmatter.com/

Pipe Properties
http://www.processassociates.com/process/basics/pipeprop.htm

Single Phase Pressure Drop in Piping
http://www.processassociates.com/process/fluid/dp_1.htm

SI Units
http://physics.nist.gov/cuu/Units/index.html

A Technical Glossary
http://flw.com/glossary/glossary_a.php

The Thermodynamic Properties of Steam
http://www.cstl.nist.gov/div836/836.01/PDFs/1955/Circular_564_H2O.pdf

Volume Flowrate Conversion Calculator
http://www.processassociates.com/process/convert/cf_flv.htm

Index

acceleration 11, 45, 100–103, 105, 109, 127, 157, 172, 183–185, 194, 218, 228, 256–259, 262, 302
accelerometers 6–7, 11, 47, 98–104, 127, 138-140, 157, 218, 253, 257, 262, 305, 307–308
accuracy requirements 304, 310
air buoyancy 186, 242
alias signal 151
alpha coefficient 62, 298
aluminum 214
amplifier 7, 47, 73, 77, 102, 118–127, 132, 160, 165, 218, 255, 260, 262
analog to digital converter 42, 130, 148–149
anti-aliasing filter 151
asset number 172–173, 176
ASTM 229
atmospheric pressure 68–70, 215
axial misalignment 234

band pass filter 128
bandwidth 120–122, 125
bar graphs 8
bees wax 103
bell prover 241
bend radius 135, 139–140
blackbody 66
bluff body 83, 240
boiling point 55, 215–216
bourdon tube 3, 69

calibration 2, 6, 20, 22–23, 25–26, 29–33, 37–43, 58, 62, 82, 123, 125, 127, 141, 153, 159–161, 163–183, 188–189, 193, 197–198, 200–204, 206, 215–223, 225, 227–229, 231, 233–236, 239–247, 249, 251, 253, 262–267, 269–270, 278–280, 286–289, 293, 296, 300, 309–310, 314
calibration cycle 25, 27, 30
calibration laboratory 163, 165–166, 172, 181, 197, 203, 227
calibration period 162, 169, 179
calibration procedures 40, 174–175
calibration records 2
calibration system software 163, 169, 172, 175
calipers 249
capacitance 74, 76, 91, 93, 127, 136, 168, 217
certificate of calibration 179
certificate of compliance 179
charge amplifier 7, 126–127
charge mode transducers 74
clamp-on ammeters 219
Class I interface 17
Class II interface 18
cold junction 207, 223
compression 105, 107, 140, 228, 233–234
compressive 109
concentric, sharp edged orifice 79
confidence 12, 160, 236, 271, 283–284
conformity 39
constant voltage 120
conversion rate 131
coordinate measuring machines 249
corner frequency 49–50, 128–129, 151
correction factor 280–281
coupling fluid 85
coverage factor 283–284
critical flow venturi 246
critically damped 50

current 4, 6, 13, 50, 61, 65, 76, 120, 124–125, 129, 135, 137, 168–169, 173–174, 217–219, 222–224, 264, 307

data acquisition 8–9, 33, 47, 74, 122, 125, 130, 143, 146–151, 156, 158, 174, 217, 231, 241–244, 251, 261, 297, 305, 307, 310
data acquisition system 8, 33, 130, 143, 147–149, 151, 158, 175, 231, 241–244, 251, 307
data analysis 12, 47, 128, 158, 307–310
data analyzers 9
data loggers 8, 146
dead weight tester 168, 177, 181, 184–186, 188–191, 296
deadband 29–31, 41
decibels 48, 128, 253
density 71, 77, 85, 186, 189, 195, 228, 239, 241, 246, 312
dew point 228, 301
dial indicators 249, 251
differential input 132, 149
differentiation 157
digital indicator 42, 143
direct memory access 148
displacement 2, 14, 89, 91, 93–95, 97, 99, 104, 114, 127, 140, 154, 157, 173, 218, 241, 243, 247, 249–251, 253, 256–259, 262, 298, 302
displacement transducers 95
display 1, 3, 7–9, 42, 47, 60, 62, 117–119, 121–122, 125, 128–130, 143–145, 147, 156–158, 165–167, 169, 174, 193, 197, 200, 203, 207, 223, 225, 229, 235, 251, 264, 269, 276–277, 307, 310
distilled water 189, 215–216
Doppler flow meter 84
Doppler shift 84
dry block calibrators 198
dry well calibrators 203
dynamic gravimetric system 244
dynamic response 46–47, 49–50

eccentric orifice plate 79
EEPROM 16
electrodynamic velocity transducers 100
electro-magnetic 75
electromagnetic interference 117
emissivity 66–67, 205
equalizing block 202–203
equalizing manifold 287
excitation voltage 6–7, 76–77, 114, 118, 124–125, 135, 137, 148, 156, 169, 229, 264, 266
expanded uncertainty 283

fixed point 212–214
flash point 201
flow 2, 11, 14, 42, 53, 70, 72, 77–81, 83–87, 129, 132, 136–138, 143, 153, 155, 159, 162, 191, 200, 202, 212, 239–247, 262, 279, 287, 293, 304, 306–308
flow calibration 239–240, 244
flow nozzle 80, 240
flow profile 78, 85, 87, 247
fluidized bed 202
force 2, 6, 11, 68, 71, 73, 97, 100–103, 105–106, 109, 141, 172, 183–185, 191–192, 194, 227–229, 231, 234, 251, 319
forward traceability 174
Fourier 9, 158
freezing point 55, 213–214
frequency 9, 12, 15, 47–50, 75, 83–85, 87, 89, 93, 100, 102–104, 117, 122, 125, 127–129, 136, 138–139, 146, 148, 151, 154, 157–158, 168, 174, 192, 217–219, 225, 240, 253–255, 257, 260, 262, 285, 297–298, 302, 305–306, 308
frequency domain 158
frequency response 12, 15, 47–50, 100, 103, 136, 285, 297–298, 305–306
full bridge 114
fume hood 201, 203
fundamental frequency 157

gauge block 250
gauge factor 109, 156, 266–267
gauge pressure 68–70, 193, 297
gallium 213
gas prover 241
gravity 71, 97–98, 100, 105, 172, 184–186, 194, 228
ground 5, 129, 131–132, 137–138, 150, 186, 250
ground loops 129, 137
grounding 138

half bridge circuit 114
harmonic distortion 262
height gauges 249
high pass filter 73, 102, 127–128, 255
hydraulic jack 230–231
hysteresis 26–27, 29–32, 40–41, 189, 191, 277, 289–290, 300

ice bath 210, 212, 214–215
ideal voltage source 120
IEEE 1451.4 16
IEEE-488 225
IEPE transducers 17
immersion depth 138, 209

inclinometer 97
independent linearity 33
indium 214
infrared 66, 205
in-line amplifier 75
input impedance 120–121, 264
instrumentation 1–2, 14, 138, 159, 161–162, 164, 168, 177, 183, 218–219, 303–306, 309–310
integrated circuit temperature transducer 65
integration 131, 135, 150, 157
integrators 99–100, 262
interferometer 250
International Organization for Standardization 244, 270, 279
international standard 176
international temperature scale 208, 210, 214
intrinsic standard 178
isothermal block 202
ITS-90 208, 210, 214

kilopascal 71, 188

laminar 78
laser Doppler anemometer 87
laser interferometer 171, 250, 262
linear encoder 241
linear variable differential transformer 89
linearity 12, 20, 23, 32–34, 36–39, 41, 160, 189, 192, 264–265, 277, 289–290, 299–300
live zero 219
load button 108
load cell 7, 106–108, 140, 227, 229–231, 233–234, 263
load transducer 105
load washers 108
local gravity 186
low pass filter 102, 127–128, 151, 255
lower range limit 19
LVDT 14, 89–90, 95

magnetic flux 76, 89
magnitude ratio 48
manifold 287
manometer 193–195
mantle 211–212
mass 10, 12–13, 46, 65, 70, 77, 84, 97–98, 101–106, 136, 139, 153, 167, 176, 183–185, 189, 193–195, 227–228, 239, 244, 246
mass flow 71, 77, 246
measurand 279–282
measurement equation 280–283

measurement uncertainty 12, 168, 279–280, 284–285, 310
megapascal 71
mercury 214
micrometers 93, 249
microstrains 267
mini-baths 198
moisture 43, 60, 127, 135
molten salt 201, 203
motion 2–3, 45, 53, 76, 94, 103, 128, 138, 157, 211, 218, 244, 253, 255, 262
multi-function calibrator 168, 178
multiplexer 8, 148–149

national standard 177
national standards laboratory 177
negative temperature coefficient 63
Newton 71, 106
NIST 177, 215, 227, 229, 239, 270, 279–280, 284, 311, 314
noise 7, 73, 75, 101, 117, 120–121, 125, 128, 133, 136–138, 146, 151, 155–157, 206
non-conformance 175
normal distribution 155–156, 270, 272, 275, 284, 299
notch filter 128
NRC 279, 311
null position 90
Nyquist theorem 151, 305

operating principle 6–7, 12, 72, 287, 298
optical coupling 129
orifice plate 79–80, 82–83, 240, 287
oscilloscope 8–9, 157
output impedance 120–121, 126
overshoot 25, 50–51, 104

parallax 195, 269
Pascal 71
peak to peak 255, 257, 259
permanent pressure loss 80
phase response 129, 255
phase sensitive demodulator 90
phase shift 48–49, 90, 157, 250, 255
piezoelectric 6–7, 47, 76, 101, 103, 126–127, 253, 298
piezoelectric accelerometer 101
piston 184–190, 227, 234, 241–242, 262–263
piston prover 241
platinum 60–62, 167, 177, 208, 222
positive temperature coefficient 63
power amplifier 218, 261

pressure 2–3, 5, 7, 11, 13, 19–20, 26–27, 30, 32, 40, 42–43, 45, 47–48, 55, 68–80, 83, 85, 101, 103, 109, 118, 122, 126–127, 135, 138–140, 143, 146, 153, 156, 160, 162, 164, 168, 170, 172–173, 177, 183, 185–188, 190–195, 197, 202, 212–216, 219, 228, 230, 234, 239–244, 246, 262–263, 279–280, 285–289, 291–294, 296–301, 305–307, 309
primary axis 140
primary standard 177, 185, 190, 193, 209–210, 239
process instrumentation 15
prover 241–243
proving ring 227, 233, 263
PVTt 240, 245

quadrant edged orifice plate 79
quality assurance system 175
quarter bridge circuit 114

radiation 5, 7, 11, 13, 135–136, 139, 285, 306
radio frequency interference 75, 117
range 6, 9, 11, 15, 19–23, 27, 30, 32–34, 36–37, 42, 47–48, 53–54, 57–58, 63, 65–66, 72, 77, 79, 82–83, 89, 93, 96, 98, 100, 118, 122, 124, 126–128, 131, 139, 144–145, 150, 155, 157, 160, 162, 168, 171, 178–179, 183, 186, 188, 191–192, 195, 198, 201–208, 214, 218–224, 227, 229, 233, 236, 239, 241–243, 245, 251, 253–255, 257–260, 262, 284–285, 287, 289, 296, 298–299, 301–302, 304–305
rangeability 11, 79, 83, 287
recorder 125, 146, 157
re-entrant well 211–213
reference standard 177–178, 253–255
regulatory requirements 304, 309
relative humidity 135, 170–171, 301
repeatability 40–41, 167, 192, 277
resistance 4, 6, 13, 39, 46, 60–63, 72, 74–75, 94, 96–97, 109, 111, 113, 125, 136–138, 156, 160, 165, 167–168, 206, 208–210, 217–218, 220–222, 235, 263, 265–267, 280–281, 285, 298
resistance bridges 222
resistance strain gauge 109
resistance temperature detectors 6, 46
resolution 42, 96, 131, 145, 149, 176, 200, 206–207, 218, 220–221, 224, 250–251, 298
resonant frequencies 12, 262, 302
resonant frequency 102–104, 254, 262
resonant peak 254
response time 49–50, 63, 138, 297
reverse traceability 174
Reynolds number 78

rise time 49
RMS 218, 256
root mean square 218
rosettes 111, 157
rotational variable differential transformer 95
RS-232 148, 225
RTD_ 6, 39, 46, 60–63, 65–66, 203, 206, 208, 220–222
RVDT 95, 97

sampling frequency 151, 305
secondary standard 200, 204
segmental orifice plate 79
self heating 13, 222
sensor 3, 6, 9–12, 29, 42, 63, 65–66, 74, 84, 89, 97, 119, 197, 200–204, 206, 222, 240, 243, 249–250, 255
serial interface 16
shaker table 218, 260, 262
shield 75, 137–138
shielding 121, 135–137
shock 122, 135, 150, 164
shunt calibration 126, 236, 263
side load 140, 302
side loading 302
signal conditioner 3, 7, 60, 89–90, 118, 130, 132, 136, 143, 165, 229, 235, 276–277
signal conditioning 5–8, 10, 15, 20, 45, 47, 60–62, 74–77, 82, 87, 90, 94, 102–104, 117–119, 122, 128–129, 131–132, 135, 143, 147–150, 153, 197, 206, 222, 255, 262, 264, 310
silver 214
single ended 107, 131, 133, 149
smart transducers 16
software validation 174
span 19–20, 23, 29–30, 41–42, 153, 160, 165, 264, 280, 286–287, 289–290, 292–293, 296, 299
specification sheet 12–13, 19, 43, 135, 286, 288, 298–299
specifications 2, 6–7, 13, 19, 35, 43, 46, 122, 135–136, 175, 179–180, 198, 203, 218–219, 254, 275, 280, 285–289, 297–301, 309
SPRT 62, 167, 177, 181, 200, 205, 208–210, 214
stability 161, 167, 169–171, 179, 198, 202–203, 205, 208, 213, 295–296
standard deviation 275, 281–283
standard gravity 186
standard platinum resistance thermometers 62, 208, 222
static 15, 19, 288–290, 292, 294–296
static electricity 171
static pressure 287, 289, 293, 296

Steinhart-Hart equation 63
stem conduction 13, 204, 209
strain 2, 4, 6–7, 26, 43, 72–73, 75, 83, 100, 103, 105, 109–111, 113–114, 118–119, 122–125, 128, 132–133, 137, 140–141, 150, 156, 235, 263–264, 266–267, 298, 307, 314, 319
strain gauge accelerometer 103
strain gauges 4, 6–7
stress 26, 105, 109, 111, 113, 126, 156, 263–264
system calibration 126, 229

TAR 170, 179, 221
tare 228
TEDS 16–17
temperature 2, 5, 7, 11, 13, 39, 43, 46, 53–56, 58, 60–66, 71–77, 79, 84, 103, 107, 113, 117, 121, 124, 135, 138–140, 143, 153, 156, 162, 164–166, 170–172, 177, 179, 186, 192, 195, 197–198, 200–204, 206–216, 218, 220–223, 228, 239, 241–244, 246, 249, 264, 279–281, 285, 287–288, 290, 293, 296–298, 300–301, 306, 308
temperature bath 165–166, 198
temperature transducers 60
temperature uniformity 198
templates 17
tensile 109
tension 105, 107, 140, 186, 228, 233–234
terminal based linearity 33
test accuracy ratio 170, 179, 222, 278
test measurement instrumentation 15
test objective 304–305, 310
test procedure 2, 303, 307, 309–310
test uncertainty ratio 170, 278
thermal expansion coefficient 249
thermal flow meter 84
thermistors 39, 63, 66, 83, 203, 220
tilt sensor 97
time constant 50, 73, 127
time domain 9, 157–158, 308
tin 214
tolerance 160, 162, 234
torque wrench 263
TPW 210–212
traceable 175, 179, 181, 240, 270, 279
transducer 3, 6–7, 10, 13–14, 19–23, 26, 29–33, 39–43, 45, 47–51, 54, 56, 65, 69–77, 81, 85, 89, 93–94, 96–97, 100–103, 105–108, 117–118, 120, 122, 124–128, 130, 132–133, 135–136, 138–141, 143–145, 147–148, 153–154, 160, 165–169, 173, 189, 193, 217–218, 228–229, 235–236, 240, 246, 249–251, 253–255, 260, 262–263, 270, 276–279, 285–287, 297–302, 305–307, 309–310

transfer standard 177–178
transformer coupling 129
traveling standard 178
traverse 78
triangulation 93, 250
triboelectric effect 75
triple point 55, 210–212, 214
triple point of water 55, 210–212, 214
TUR 170, 173, 175, 179
turbine flow meter 42, 82
turbulent 78

ultrasonic displacement transducer 94
unbroken chain 270, 279
uncertainty 12, 14, 198, 212, 215, 227, 242–243, 279–283, 285, 298, 310
underdamped 50
upper range limit 19, 22

vacuum 5, 68, 176, 183, 186, 246, 297, 301
variable reluctance 76, 103–104
variable reluctance accelerometer 103
variance 272–275, 281–283
velocity 45, 54, 77–78, 84–87, 94, 99–100, 105, 109, 127, 154, 157, 218, 253, 256–259, 262, 302
velocity transducers 99–100
venturi 80, 240, 246
vibrating reed 233
vibration 2, 5, 9, 11, 14, 43, 61, 89, 99, 101–102, 135–136, 139, 143, 150–151, 154–157, 159, 162, 164, 171–172, 179, 218–219, 253, 255–257, 259–260, 262, 285, 300–302, 306, 308
voltage 4, 6–7, 15, 42, 56, 60–61, 65, 74, 76, 89, 91, 93, 104, 114, 118–126, 130–133, 137, 143, 153, 165, 168, 174, 205, 217–220, 222–224, 229, 264–266, 280
voltage amplifier 119–120
volumetric flow 70, 77, 82, 239, 245, 247
vortex flow meter 83
vortex shedding 83
vortices 83, 307

weigh tank 243–244
Wheatstone Bridge 6–7, 61, 72, 75
working standard 168, 186, 190, 221

zero based linearity 33
zinc 214